PREMODERN TRANSLATION

CONTACT AND TRANSMISSION

INTERCULTURAL ENCOUNTERS FROM
LATE ANTIQUITY TO THE EARLY MODERN PERIOD

VOLUME 2

Premodern Translation

Comparative Approaches to Cross-Cultural Transformations

Edited by

SONJA BRENTJES AND

ALEXANDER FIDORA

BREPOLS

Cover image: 'Jacques de Guyse at his desk with books', in the Middle French translation of his *Annals of Hainault* by Jean Lebasse; painting attributed to the Master of Antoine Rollin. Oxford, Bodleian Library, MS. Douce 205, fol. 2v. *c.* 1500. Reproduced with permission.

D/2021/0095/18
ISBN 978-2-503-59097-4
eISBN 978-2-503-59098-1
DOI 10.1484/M.CAT-EB.5.121169

Printed in the EU on acid-free paper.

Table of Contents

SONJA BRENTJES AND ALEXANDER FIDORA

Introduction

The papers in this volume explore possibilities for contextualizing and questioning the well-established histories of translation in medieval societies. One author asks traditional questions about the technologies of translation for a far-away culture, in the hope of enabling readers familiar with translations into Latin or Hebrew to reflect on shared cross-cultural practices. Other authors ask questions of a less traditional nature about the character or role of submerged translations that never appear in any of the traditional histories of translation in medieval societies, with regard to the mathematical, medical, or philosophical sciences. Finally, a third group of authors deals with topics at best rarely discussed in relationship to translations, thus providing novel perspectives from outside the traditional domain of research.

Important interpretative advantages of disciplinary border crossings regarding narratives of translation told by historians of science, medicine, or philosophy for the eighth to the thirteenth centuries in the Islamicate and Catholic domains of Europe, North Africa, and West Asia arise primarily from the diversity of historical conditions that formed the practices, products, institutions, and economic as well as sociocultural opportunities in each individual phase. This diversity encourages researchers to pose new questions and focus on aspects of medieval sciences in Christian or Islamicate societies hitherto neglected; these approaches enable us to learn about the materials which created the stepping stones for telling modified or altogether new narratives about translation. This is, indeed, also the case for the six papers brought together in this volume.

Yukiyo Kasai's survey of Old Uyghur translations of Tocharian and Chinese Buddhist texts leaves no doubt that the technology of translations was both common to all medieval cultures engaged in translation, and culturally specific. Their commonality resulted from the basic task of rendering a text and its content in a new language. This task was achieved by following one of two basic patterns: either translating the smallest possible philological units into an equivalent, or translating larger philological units as a whole. The

Sonja Brentjes • (brentjes@mpiwg-berlin.mpg.de) is a retired historian of science in Islamicate societies and Christian Europe.

Alexander Fidora • (alexander.fidora@icrea.cat) is an ICREA research professor in the Department of Ancient and Medieval Studies of the Universitat Autònoma de Barcelona.

cultural specificity reflects the translators' skills and tools, their audiences' expectations, and their society's level of cultural complexity, as well as the nature of the texts — whether they had already undergone the process of translation or not. In the case of translations into Old Uyghur, different source languages encouraged different translational practices, while different formats of preserved translations seemed to reflect different functional contexts and reading practices. These are aspects that have been underappreciated in the major research areas of translating scholarly works in western Asia, Sicily, the Iberian Peninsula, or southern France.

On the basis of two Latin mathematical texts, the *Liber mahameleth* and Fibonacci's *Liber abbaci*, Jens Høyrup argues for the existence of hidden stories of translation, which the authors of mathematical texts, or perhaps even the translators who preceded them, had concealed and covered up. Against the humanist story of a rejection of translations from Arabic and their subsequent modification into a Latin filled with Greek words and grammatical constructions — the most extreme example of which known so far seems to be the *Liber de canonio* — Høyrup's examples include, at least in one case, a hidden translation that may once have been made of a Greek text. While he does not establish why some translations of mathematical works were presented as 'non-translations', he shows very clearly that problems, methods, or solutions which had been divested of their identities as translations, as well as their itineraries, can only be rediscovered through an analysis of their technicalities. From this perspective, non-translations — and possibly even their underdeveloped narratives — have no context except the technical.

The chapters by Alexander Fidora, Katja Krause, and Henryk Anzulewicz question how useful it is to separate translation from other knowledge practices such as transmission, dissemination, or transformation. Fidora focuses on Dominicus Gundissalinus, who was one of the most prolific translators of philosophical texts from Arabic into Latin during the twelfth century, as well as one of the first authors to systematize the new philosophical doctrines. A very conspicuous example of this effort to systematize the doctrines that he himself had made available through his translations is his *De divisione philosophiae*, a treatise which conferred Gundissalinus's translations a lasting influence on scholastic philosophy. The division of the sciences, which Gundissalinus drew from his Arabic sources, was crucial not only for the development of specifically epistemological discussions, but also crossed genres and became a key ingredient of Aristotelian commentary literature. Krause and Anzulewicz explore the concept of interpretation in the works of Albertus Magnus, arguing that our understanding of such terms is radically different from that of medieval scholars, such as Albert. Such differences need to be taken seriously if we wish to gain an understanding of translations and their consequences across languages, epistemes, experiences, and theologies, or moral economies. They show that the multiple acts of translations that took place from Late Antiquity to the thirteenth century between different languages and scholarly systems provided Albert with different scholastic libraries and, thus, various intellectual choices and opportunities for interpretation through rearrangements, reformulations, and extensions. This multiplicity of acts of translating did not simply lead to linear, uniform cultural choices, as is often claimed when discussing translations from Arabic or Greek into Latin during the twelfth and thirteenth centuries, but enabled Albert to formulate new ideas, reshape previous disciplinary affiliations, and build a novel system of knowledge about God, nature, the intellect, and human happiness.

Historiographically closely related to Fidora's, Krause's, and Anzulewicz's position, the overarching concept of this anthology aims to broaden our perspectives on translating as an important tool in cultures, which has enabled and promoted the rise of scholarly activities and communities incorporating at least two, but usually multiple, languages and epistemes. Reducing translations for and within such communities to scholarly texts and practices alone, as in much academic research to date, has isolated them from other translational activities in the same time and space. Translation has become separated from the other knowledge practices discussed by Fidora, Krause, and Anzulewicz, and the focus on elite languages and knowledge forms has impoverished our understanding of the complexity under which such specialized practices of translation took place. Communities translating into or from so-called vernaculars or privileging lower registers of knowledge, such as divination or popular magic, were often ignored until the late twentieth century, in the same manner as were historical, religious, legal, diplomatic, or economic documents and their translational and other contexts.

Claire Gilbert picks up on one strand of such 'outsider' activities in a time when at least one of the vernaculars on the Iberian Peninsula, Castilian, gained higher political and cultural prominence within and outside of the Kingdom of Castile and thus experienced a new, more rapid phase of formalization and codification, while a previously dominant elite language of knowledge, Arabic, together with its vernacular registers, began a steep downturn. She highlights the different registers of Arabic that existed and were used in diverse circumstances in sixteenth-century Castile, and thus needed translators with often very different skills and hence training. Inquisition translators, for instance, did not think of their work as translating but as characterizing or describing specific types of texts, while frontier translators needed to guarantee the provisioning of their garrisons by using locally-spoken variants of vernacular languages. Translators of Arabic legal documents at courts, in contrast, had to be able to deal with a highly specialized formal language and produce formalized written outputs that could withstand a series of legal challenges. The main body of Gilbert's contribution focuses on the continuous learning and honing of a variety of linguistic skills by one such Spanish translator of Arabic, the *morisco* Alonso del Castillo. Analysing marginal notes in one of the surviving manuscripts from Alonso's library, Gilbert claims that the particular usefulness of a Spanish Arabic translator consisted in the limitations of his linguistic knowledge, his willingness and ability to learn and, thus, to diversify those limitations. Such a view of a translator's skills seems to be counterintuitive when seen from the perspective of standard narratives of medieval scientific, philosophical, medical, and related translations, where at best three or, recently, also four linguistic levels are admitted — scholarly Arabic and Latin and some vernacular registers — and limitations are considered obstacles that need to be overcome.

In fact, in early modern Spain and its Moroccan diplomacy, such limitations created points of contact or even interaction between two scholarly domains: the long-established lexicographic tradition of Classical Arabic and the newly-emerging lexicography of vernacular Castilian. As a consequence, this volume shows that, where Albertus Magnus interpreted through rearrangements, reshaping of affiliations, or novel ideas and structuring, Alonso del Castillo translated through omission, reformulation, explanation, and novel composition. Similarly, while Høyrup suggests restoring the fate of 'non-translations' to the spectrum of academic narratives on translations through in-depth technical analysis, Gilbert proposes telling narratives through a painstaking investigation of the traces people left in the margins, because — at least in her case — these reveal that their actors were rather close to the centres of overlapping and competing translational, and even broader intellectual, trends.

In the final chapter, Pietro Omodeo goes one step further. He unearths an additional layer of consequences that translations have had on mathematics and astronomy, through the modes in which the narratives on translations underpinned the two disciplines. He explores how those narratives changed during the century that separated two of the disciplines' leading experts — Johannes Regiomontanus in the fifteenth century and Erasmus Reinhold in the sixteenth century. Like the previous authors, Omodeo reflects on the relevance of methodological choices when interpreting narratives and, like Krause, Anzulewicz, and Gilbert, he draws on a rich and diversified debate about how to analyse concepts, rhetorical styles, or entire stories. While Regiomontanus told his story in the traditional style of the so-called eulogies of the arts, his cultural universe was multidimensional, conceptually governed by language and, thus, plurilingual. In his era, five language cultures dominated the entirety of human knowledge — Greek, Latin, Arabic, Persian, and Indian. The transfer of knowledge between them was enabled through translation and unfolded via appropriation, elaboration, and new syntheses. Other cultures like those of the Egyptians or the Chaldaeans had their place too, primarily in the origin phases of various scholarly disciplines. But Regiomontanus's outlook went beyond praising previous intellectual cultures, their languages, and the crucial need to translate in his own time. Omodeo shows that Regiomontanus's narrative of translations and knowledge interweaved a view of past perfection in morality, eloquence, and intellectual competence with a biting, although already topical critique of the mercantile values and academic decadence of his own times, in order to legitimize an ambitious programme of translation, publication, and curricular reform. Despite its often traditional nature, Regiomontanus's approach revealed his own deeply political and sociocultural outlook, which was subsequently implemented by his students and admirers.

Reinhold's eulogy of Regiomontanus served to congratulate the first students of the reformed mathematical curriculum at the University of Wittenberg. It embodied, as Omodeo argues, a profound shift in historical, moral, and pious outlook. The multiplicity of languages and knowledge cultures was sacrificed in favour of ancient, primarily Greek, authors and their works, as well as apocryphal or even invented biblical traditions. University reform was replaced by spiritual reformation, or, as Omodeo concludes, the Renaissance and its moral economy had turned into the Reformation with its values of studying the heavens for the higher good of knowing God and His Deeds. Translation as a multilingual, cross-cultural activity had been replaced by a practice of a showy bilingualism. Religious and academic politics had come to the forefront of the previously primarily cultural legitimation of translation for the sake of knowledge rejuvenation, educational reform, and as a way to address new intellectual challenges.

The six chapters united in this volume highlight the diverse options available for dynamizing and expanding the still all-too-often linear and static interpretations of sociocultural processes of translation and the cross-cultural spread of knowledge between the eighth and the thirteenth centuries, when comparatively situating them within broader and longer histories of translations and their consequences in the premodern and early modern world.

Studies of translational methods and technicalities, as important as they are for determining linguistic options or tools, translators' habits, or hidden audiences, do not suffice for determining functions, purposes, or values of translations. As Bruce O'Brien has shown, with regard to translation cultures in medieval England, and as the authors of the six papers in this volume argue in their own respective ways, it is vital to pay attention to reading habits, educational opportunities, epistemic goals, moral values, economic prospects, or political contexts in order to go beyond the limits of our traditional narratives of movements and Greek prevalence, to create historical interpretations that allow for greater depth, multiplicity of practices and perspectives, as well as locating them in specific knowledge cultures and their societies.[1]

1 Bruce R. O'Brien, *Reversing Babel: Translation among the English during an Age of Conquests, c. 800 to c. 1200* (Newark: University of Delaware Press, 2011).

YUKIYO KASAI

Old Uyghur Translations of Buddhist Texts and Their Usage

Introduction

In the middle of the eighth century, the Uyghurs, a Turkic-speaking tribe, founded their first nomadic Empire in Mongolia and exerted a dominating influence not only there, but also more widely across Eastern and Central Asia.[1] This powerful empire, however, declined in the middle of the ninth century and a large part of the Uyghurs migrated in a westerly direction, to the eastern part of the Tianshan 天山 area.[2] There they established their new kingdom, the 'West Uyghur Kingdom', which centred around the Turfan area which today lies in the Xinjiang Uyghur Autonomous Region of the People's Republic of China. When, in the thirteenth century, the Mongols expanded their power in Central Asia under the rule of Chingiz Khan, the Uyghur king decided to submit voluntarily to this new rising power. Because of this decision the Uyghurs were favoured under Mongol rule, and their kingdom enjoyed a certain freedom for a while.[3] So, until the Mongol Empire moved its centre northwards in 1368 CE, the Uyghurs played an important role as rulers or rulers' advisors in the history

1 The history of the Uyghurs and their first nomadic empire were dealt with by many scholars. For an overview, see e.g. Mackerras, *The Uighurs*; Golden, *An Introduction to the History of the Turkic People*, pp. 155–63; Sinor, Geng and Kychanov, 'The Uighurs, the Kyrgyz and the Tangut'.

2 For their migration into the eastern part of Tianshan 天山 area, see e.g. Moriyasu, ウィグルの西遷について [Nouvel examen de la migration des Ouïgours au milieu du IXe siècle]; Moriyasu, *Die Geschichte des uigurischen Manichäismus an der Seidenstraße*, pp. 165–66.

3 On the voluntary submission of the Uyghurs and the development of their situation, see e.g. Saguchi, モンゴル人支配時代のウイグリスタン (上) (下) [Uiguristan under the Mongol rule (I, II)]; Abe, 西ウイグル国史の研究 *Nishi Uigurukokushi no kenkyū*, pp. 1–138; Allsen, 'The Yüan Dynasty and the Uighurs of Turfan in the 13th Century', pp. 246–80.

Yukiyo Kasai • (yukiyo.kasai@ruhr-uni-bochum.de) is a researcher of Old Uyghur philology and history. She is working in the ERC project 'BuddhistRoad' housed at the Center for Religious Studies at the Ruhr-Universität Bochum.

Premodern Translation: Comparative Approaches to Cross-Cultural Transformations, ed. by Sonja Brentjes and Alexander Fidora, Contact and Transmission, 2 (Turnhout: Brepols, 2021), pp. 13–31

10.1484/M.CAT-EB.5.122132

of Central Asia. Under Mongol rule the Uyghurs flourished — not only in the political arena, but also in the economic, military, cultural, and religious spheres. It is certain that Buddhists occupied a key position in these diverse activities.[4]

During the time of the nomadic empire, the Uyghurs accepted Manichaeism, a religion that had been founded by Mani in third-century Babylonia, and the rulers gave it the royal support continuously.[5] After the fall of this empire and its people's migration in a western direction, they remained Manichaeans for a while.[6] Because of the influence of the local inhabitants in the Turfan area, mainly Tocharians and Chinese, however, the Uyghurs gradually became Buddhist. In the introduction of Buddhism, these local ethnic groups played an important role, but over time the Chinese exerted a stronger influence and finally became the Uyghurs' main source of Buddhism.[7] In the eleventh century Buddhism obtained the position of the dominant religion in the West Uyghur Kingdom, remaining as such after the establishment of the Mongol Empire. It is also known that the Uyghurs' Buddhist worship even influenced their new rulers, the Mongols.[8]

---

4 The activities of the Uyghurs under Mongolian rule are, for example, dealt with in M. Brose's monograph, *Subjects and Masters*. Other relevant researches on this topic are gathered and summarised in Kasai, *Die uigurischen buddhistischen Kolophone*, pp. 15–19.

5 The church history of Manichaeism and its teaching have been dealt with in many books and articles. For a short introduction, see e.g. Tardieu, *Manichaeism*. The conversion of the Uyghurs to Manichaeism is one of the well-discussed topics, see e.g. Chavannes and Pelliot, 'Un traité manichéen retrouvé en Chine', pp. 190–99; Clark, 'The Conversion of Bügü Khan to Manichaeism'; Moriyasu, *Die Geschichte des uigurischen Manichäismus an der Seidenstraße*, pp. 34–35.

6 Thus, a small number of the Old Uyghur Manichaean texts was found in the Turfan area, too.

7 See e.g. Moriyasu, *Die Geschichte des uigurischen Manichäismus an der Seidenstraße*, pp. 35–38, 174–92; Tremblay, 'The Spread of Buddhism in Serindia-Buddhism Among Iranians, Tocharians and Turks Before the 13th Century'. The role of Sogdian Buddhism amongst the Uyghurs was also discussed. There is the other hypothesis that the Sogdians already introduced Buddhism to the Turks under the rule of the Tujue 突厥, the predecessor of the Uyghurs (Sogdian hypothesis), see e.g. von Gabain, 'Buddhistische Türkenmission'; Laut, *Der frühe türkische Buddhismus und seine literarischen Denkmäler*, pp. 1–12. Considering the fact that numerous Sanskrit Buddhist terminologies were borrowed via Tocharian into Old Uyghur, the strong influence of the Tocharians is obvious, see e.g. Mironov, 'Kuchean Studies', pp. 158–68; Shōgaito, '古代ウイグル語'におけるインド来源借用語彙の導入経路について (On the Routes of the Loan Words of Indic Origin in the Old Uigur Language)'; Moerloose, 'Sanskrit Loan Words in Uighur'. The scholars who support the 'Sogdian hypothesis' recognize the important role of the Tocharians on the Uyghurs' Buddhist worship, too, see e.g. Geng, Laut and Pinault, 'Neue Ergebnisse der Maitrisimit-Forschung (I)', 'Neue Ergebnisse der Maitrisimit-Forschung (II)'. Furthermore, according to Y. Yoshida's recent research the Uyghurs seem to have used the Sogdian Buddhist sutras, at least, see Yoshida, 'Die buddhistischen sogdischen Texte in der Berliner Turfansammlung', pp. 340–44. Thus, this topic needs further discussion.

8 The Buddhist influence of the Uyghurs on the Mongols can be seen in the fact that many Mongolian Buddhist terminologies were borrowed from Old Uyghur, see e.g. Kara, 'L'ancien ouigour dans le lexique mongol'; Shōgaito, 'Uighur Influence on Indian Words in Mongolian Buddhist Texts'.

During the West Uyghur Kingdom era and, subsequently, the Mongol period, the Uyghurs produced numerous texts with diverse contents in their own language, Old Uyghur.[9] The handwritten and printed Buddhist texts, including sutras and commentaries, make up the majority of Old Uyghur textual remains. Although these are often only preserved in fragments, they emphasize the importance of Buddhism for Uyghur society at that time.[10]

The Characteristics of Old Uyghur Buddhist Texts

Considering the introduction process of Buddhism, it is no wonder that most of the Old Uyghur Buddhist texts, which were produced from the tenth to the fourteenth century, are translations from other languages. While some from the early period were translated from Tocharian, an Indo-European language, Chinese remained the main source language throughout the entire period. Besides those two languages, Tibetan also played an important role, especially in the Mongol period, from the thirteenth to the fourteenth century.

A listing of extant Old Uyghur Buddhist literature shows the wide variety of texts that were produced — not just the major sutras,[11] but also commentaries, recondite Abhidharmas,[12] and Chinese apocryphal sutras.[13] Based on those diverse texts we might be tempted to suggest the existence of an Old Uyghur canon once reported by Guiseppe Tucci as a hearsay statement from a Tibetan monk, although no such concrete canon has yet been found.[14]

9 Some of these texts survived and have been discovered through archaeological excavations in Central Asia which have taken place from the beginning of the twentieth century. The German archaeological expeditions are particularly notable for their intensive excavations in the Turfan area, which was the centre of the West Uyghur Kingdom. Because of that, numerous Old Uyghur texts are now preserved in Berlin as a part of the Berlin Turfan Collection. The images of those fragments are all available in the Digitales Turfan-Archiv: http://turfan.bbaw.de/dta/index.html. Those in other collections in the world are partly presented in the International Dunhuang Project, see http://idp.bl.uk/.

10 J. Elverskog's monograph gives a good overview of the Old Uyghur Buddhist texts, see Elverskog, *Uygur Buddhist Literature*. After the publication of this monograph many further texts have been identified. Some of them are published in the series of *Berliner Turfantexte*. The list of published volumes is available at http://turfan.bbaw.de/publikationen.

11 Many major Buddhist sutras are listed in Elverskog, *Uygur Buddhist Literature*, pp. 17–73.

12 In Elverskog's book, several commentaries and Abhidharma texts are listed, see Elverskog, *Uygur Buddhist Literature*, pp. 74–85, n. 40–49. Further commentaries are also identified, see, e.g. Kasai, *Der alttürkische Kommentar zum 'Vimalakīrtinirdeśa-Sūtra'*. In addition, in my current work further Abhidharma texts which are partly written in Brāhmī script were identified, see Kasai, *Die altuigurischen Fragmente mit Brāhmī-Elementen*, pp. 29–88, Texte A.

13 General information of Old Uyghur version of several apocryphal sutras, see Elverskog, *Uygur Buddhist Literature*, pp. 86–104, n. 50–58.

14 Several scholars tried to investigate the existence of the Uyghur Buddhist canon according to G. Tucci's statement. This procedure is well summarized by Wilkens, see Wilkens, 'Hatten die alten Uiguren einen buddhistischen Kanon?', pp. 355–57. He himself tried to make this question clear from the aspect of philological and religion studies, but a concrete conclusion is still open.

When we take a closer look at the contents of the texts, some remarkable characteristics become apparent. Among the translations from Tocharian, the Buddhist narratives which played an important role in Tocharian Buddhism merit special attention.[15] Furthermore, the translations from Chinese contain different Buddhist tendencies, because there is such a substantial volume of them, scattered chronologically from the tenth to the fourteenth century. It is especially noteworthy that several of the texts indicate the influence of the Yogācāra school, which had been widespread since the eighth century in Dunhuang, a neighbouring oasis of Turfan.[16] Other texts show a close relationship with Buddhists in the Liao-Dynasty that had been founded by Khitan in the tenth century.[17] During the Mongol period the Uyghurs were in contact with Tibetans and, as a result of this contact, several Tantric texts were translated into Old Uyghur.[18] All of this shows that the Uyghurs were influenced by different Buddhist cultures at different times; and developed their textual traditions.

In cases where the origin of the content of Old Uyghur Buddhist texts can be successfully identified, a comparative study between the Old Uyghur versions and their possible counterparts in other languages yields interesting results.

---

15 As an example, the text *Daśakarmapathāvadānamālā* can be mentioned. The complete edition of its Old Uyghur version was published in Wilkens, *Buddhistische Erzählungen aus dem alten Zentralasien*.

16 On the introduction and the spread of the Yogācāra school in Dunhuang see e.g. Ueyama, '敦煌と曇曠の仏教学' [Dunhuang and Tankuang's Buddhism]; Ueyama, '大蕃国大徳三蔵法師沙門法成の研究' [Study on *Dafanguo dade sanzang fashi*, Facheng]. The influence of this school among the Uyghur Buddhists was first pointed out by Kudara, '妙法蓮華経玄賛のウイグル訳断片' [Old Uyghur fragment of the Miao-fa-lian-hua-jing Xuan-zan], pp. 201; and Röhrborn, 'Die alttürkische Xuanzang-Vita', p. 551. Later many facts which support their thesis were found out, see e.g. Kasai, 'Die uigurische Überlieferung der Legende von der Gründung des Tempels *Baimasi*'; Kitsudō, ウイグル語訳『観弥勒上生兜率天経賛』について [On the Uyghur version *Guan mile shangsheng doushuaitian jing zan*]; Kasai, 'Der Einfluß der Vijñānavāda-Schule'; Kitsudō, ウイグル文慈恩宗文献「大唐三藏行跡讚」 [Uyghur Yogacārā text *Datang sanzang xingji zan*]. On the Buddhist cultural exchanges between Dunhuang and Turfan, see e.g. Rong, 'The Relationship of Dunhuang with the Uighur Kingdom'.

17 Kitsudō, 'Liao Influence on Uigur Buddhism'.

18 The Old Uyghur Tantric texts were mainly investigated by G. Kara and P. Zieme, see Kara and Zieme, *Fragmente tantrischer Werke in uigurischer Übersetzung*; *Die uigurischen Übersetzungen des Guruyogas 'Tiefer* Weg'; Zieme and Kara, *'Ein uigurisches Totenbuch'*. From the historical aspect, the Uyghurs and the Tibetans may have been in contact before the Mongol period. In fact, the famous Old Uyghur text which is often mentioned under the name 'Buddhist catechism' was written in Tibetan script, see Maue and Röhrborn, 'Ein "buddhistischer Katechismus" in alttürkischer Sprache', Part 1 and 2; Moriyasu, チベット文字で書かれたウィグル文仏教教理問答 (Études sur un catéchisme bouddhique ouigour en écriture tibétaine), P.t. 1292. According to T. Moriyasu, who investigated this text in detail, the text was probably produced around the tenth century (pp. 24–25). Thus, the Uyghurs had the possibility to accept the Tibetan script within the context of Buddhist worship in that period. The influence of the Tibetan Buddhism in the pre-Mongol period still needs further investigations.

Because not all of the texts which once existed historically still exist, such an investigation has to be carried out with the assumption that the actual original has already been lost. But it is still worth remarking that many Old Uyghur translations indicate strong deviations, or rather additions, when compared to their possible original. This deviation can even be observed in the writing style. A good example of this is one of the most famous and extensive Old Uyghur Buddhist translations from the early period, *Maitrisimit*,[19] in which the future Buddha named Maitreya appears as the central figure. The Old Uyghur version was translated from the Tocharian *Maitreyasamiti-nāṭaka*.[20] While the Tocharian version was written in verse, the Old Uyghur one is in prose. However, the writing style is not the only difference between those two versions. The Tocharian version was probably used as a drama piece, so it is a relevant question whether the Old Uyghur *Maitrisimit* also fulfilled this function. Annemarie von Gabain suggested that the Old Uyghur version would have been performed as a drama on a special festival day, namely the festival of the new or full moon.[21] This idea was widely accepted[22] and Rong Xinjiang even suggested that Old Uyghur plays such as *Maitirisimit* had probably influenced the further development of folk literature in Dunhuang.[23] Von Gabain's idea was based on her interpretation of the key words *körünč* and *yaŋı kün*. While the first word was once translated as 'showpiece' or even as 'drama', the latter means 'new day'. She connected those two terms and came to the abovementioned conclusion. But Klaus Röhrborn has already pointed out that the word *körünč* can only be translated as 'wonder' when taking into account the use of this word in different Old Uyghur Buddhist texts.[24] In addition, Jens Wilkens' recent investigation shows that *yaŋı kün* was used as a common term for 'feast, festivity, ceremony' and did not have any relationship with the special festival for Maitreya.[25] Considering the new

19 For the Old Uyghur *Maitrisimit* the complete edition is still lacking. Up to now, three different manuscripts are known. The fragments belonging to two manuscripts were found in Sängim and Murtuk and are a part of the Berlin Collection. The fragments of the third one were found later than the first two from Hami and are preserved in China. For the former both manuscripts there is just a catalogue, see Laut and Wilkens, *Alttürkische Handschriften*, Part III. The fragments belonging to the third manuscript have been investigated more thoroughly than the other two versions, see e.g. Yüsüp, Qämbiri and Xoğa, 回鶻文弥勒会見記 *Huihuwen mile huijianji*; *'Das Zusammentreffen mit Maitreya'*, ed. and trans. by Geng and Klimkeit.

20 The fragments of the Maitreyasamiti-nāṭaka in Tocharian A were edited, see Ji, Winter, and Pinault, *Fragments of the Tocharian A*. The comparative study of the Tocharian and the Old Uyghur versions were also carried out, see e.g. Geng, Laut, and Pinault 'Neue Ergebnisse der Maitrisimit-Forschung', Part I and II.

21 *Maitrisimit*, ed. by von Gabain, p. 29.

22 The discussion on this topic was well summarized by Wilkens, 'Der "Neutag" und die "Maitrisimit"', pp. 376–79.

23 Rong, 'The Relationship of Dunhuang with the Uighur Kingdom', pp. 290–98.

24 Röhrborn, 'Die alttürkische "Maitrisimit"'.

25 Wilkens, 'Der "Neutag" und die "Maitrisimit"'.

interpretation of those two key terms, the theatrical use of the Old Uyghur *Maitrisimit* appears to be rather unlikely. It is difficult to understand why the Uyghur translator changed the style of the text, but the cultural background of the Uyghur Buddhists at that time surely played an important role in that decision.

In terms of the additions, it is difficult to determine whether they came from a Uyghur translator or, rather, a lost original. But some of the terminological explanations were certainly introduced by the Uyghurs. An example are the following sentences in the Old Uyghur commentary on the *Vimalakīrtinirdeśa sūtra*:

> *samar änitkäk tili ärür: türkčä yüzgärü utru kolulamak tep ter: y(e)mä üč törlüg bir yok kurug s(a)mar ikinti aṭkanguluk-suz körksüz b(ä)lgüs[üz samar] üčünč [küsä]gülüksüz samar*
>
> (*S*[*amādhi*] is Indian. In Turkish, in contrast, one says "to meditate". And there are three kinds of (*samādhis*): First, the emptiness$_2$ *samādhi* (Skt. *śūnyatā-samādhi*), second, the not-being attached, unshaped and feature[less *samādhi*] (Skt. *ānimitta-samādhi*), third, the [want]less *samādhi* (Skt. *apraṇidhāna-samādhi*))[26]

The correspondence with the possible Chinese original is quite short: *sanmei sansanmei* 三昧三三昧 'The *samādhi* is three kinds of *samādhis*'.[27] Comparing those two versions, the Old Uyghur one is clearly longer and more detailed. Because the lost Chinese version might have contained a detailed list of three kinds of *samādhis*, it is difficult to decide whether the second half of this commentary can, in fact, be recognized as an addition in the Old Uyghur version. But the first part, in which the meaning of the terminology *samādhi* is explained in Old Uyghur, was definitely added by the Uyghur author. This example shows that — at least in the translation, or rather in the revision, of the Buddhist commentaries — the Uyghur translators or authors aimed to make the context clear for their readers and were ready to add their own explanations to result in a better understanding.

Furthermore, the Uyghurs maintained the texts even after finishing their own translations and tried to make them suitable for the current conditions of Uyghur Buddhist worship. *Foshuo tiandi baying shenzhou jing* 佛說天地八陽神咒經 [*Mantrasūtra* of the Eight Principles of Heaven and Earth as Spoken by Buddha, T. 2897.85], a famous apocryphal Chinese sutra, was one of the earliest Buddhist texts to be translated into Old Uyghur, and still today there are at least two different versions, which differ considerably from

26 Kasai, *Der alttürkische Kommentar zum 'Vimalakīrtinirdeśa-Sūtra'*, pp. 108–09, ll. 295–97.

27 The Chinese counterpart can be found in T. 2777 *Jingmingjing jijie guanzhongshu* 淨名經集解關中疏, 454b28.

each other.[28] Juten Oda, who intensively investigated this text, explained that the two versions came into existence because the religious circumstances of the Uyghurs had changed. He asserted that, when it was first translated, the majority of Uyghurs were Manichaeans. Thus, it seems to have been necessary to explain some Buddhist sentences using Manichaean terminologies and teachings. But after most of them had become Buddhist, that was no longer necessary and the text's accuracy in terms of Buddhism became more important. Because of this change, this text was revised and the Manichaean elements were completely eliminated.[29] The following two translations of the same Chinese sentence show this difference clearly:

> T. 2897 *Tiandi baying shenzhou jing* 天地八陽神呪經, 1424a15.
> 皆是天之常道。自然之理。世諦之法。
>
> > (This is all the constant way of Heaven, the principle of that which is so of itself, and the law of conventional truth.)[30]

Old Uyghur version:[31]

version Ia	*bo alku yertinčü yersuvdakı ädgüli anıglı k(a)ntünüŋ iki törlüg kılınč ugrınta törütmiš törö tetir* (It is said to be a law which good and evil in this entire world system have created as a result of their own two kinds of actions.)[32]
versions Ib and IIc	*bo alku yertinčüdäki käntün törütmiš törö tetir* (It is said to be a law which those in this entire world have created themselves.)

Version Ia is the first translation, while versions Ib and IIc can be characterized as revised versions. It is obvious that neither Old Uyghur translation corresponds perfectly to the Chinese text which has come down to us. But the more serious difference between the two versions is that the underlined parts in version Ia are completely omitted in the revised versions Ib and IIc. This long phrase in version Ia mentions the contrast between good and evil,

---

28 For the general information on this sutra, see Elverskog, *Uygur Buddhist Literature*, pp. 95–100, n. 55. The complete edition with comparative study between the different versions was published by Oda, *A Study of the Buddhist Sūtra called 'Säkiz Yükmäk Yaruq*. According to his study, in each version there are still small variations.

29 See Oda, 'トルコ語本八陽経写本の系譜と宗教思想的問題 (Some Problems on Manuscripts of the *Eight Lights Sūtra* in Uygur)'. Röhrborn, however, questioned Oda's thesis, see Röhrborn, 'Zum manichäischen Einfluß im alttürkischen Buddhismus'. In response to Röhrborn's argument, Oda argued for his own thesis, again, see Oda, 'On Manichaean Expressions in the *Säkiz yükmäk yaruq*'.

30 Oda, *A Study of the Buddhist Sūtra called 'Säkiz Yükmäk Yaruq'*, p. 199.

31 Oda, *A Study of the Buddhist Sūtra called 'Säkiz Yükmäk Yaruq'*, pp. 198–99, ll. 329–30.

32 The emphasis is mine.

which played an important role in Manichaean teaching. The possibility that there once existed a Chinese version which contained exactly the same sentence cannot be completely excluded, but the striking similarity of the presented idea to the Manichaean one is worth considering.

In addition, Oda pointed out that the Old Uyghur version of the *Avalokiteśvara sūtra* had also been revised,[33] which suggests that the Uyghur Buddhists' translation activity had a certain continuity, and that the translated texts were used and carefully controlled with regard to their content.

Old Uyghur Buddhist Translations under Chinese Influence and Their Usage

As mentioned above, Chinese Buddhist texts played an important role as the main source of the Uyghur Buddhists' translation activities. The Chinese influence was so strong that there seems to have been a certain group among the Uyghur Buddhists who learned Chinese characters and their pronunciation systematically.[34] The Old Uyghur texts in which even the Chinese word 'order' was intentionally borrowed were probably produced by such a group. The Old Uyghur translation of the *Madyamāgama* can be counted as one of them:

> T. 26 *Zhong ahanjing* 中阿含經, 601a15–16.
> 若汝從如來聞大人根智分別者。必得上信如來而懷歡喜。

Hedin Collection in Stockholm 1935.52.0004^{v} ll. 3–8:[35]

birök	s(ä)n	mintirdin	äšidtiŋ	ärsär:	ulug ärän-lär-niŋ	töz tüp	bilgä bilig üzä
若	汝	從如來	聞	者	大人	根	智

bölmäk aḍırṭlamak-ıg:	otgurak	bulup	yeg üstünki	kertgünčüg:	mini-tä	inčip
分別	必	得	上	信	如來	而

öriṭgäy ärdiŋ	ögrünčüg sävinčig tep:
懷	歡喜。

> (If you heard from me the dividing and the distinguishing of the root$_{2}$ of great men through their wisdom, you would have surely found the highest belief, and nevertheless you would have aroused joy and delight in me.)

---

33 Oda, 'A Fragment of the Uighur *Avalokiteśvara-Sūtra* with Notes'.

34 Shōgaito, Masahiro 庄垣内正弘, ロシア所蔵ウイグル語文献の研究ーウイグル文字表記漢文とウイグル語仏典テキストー *Roshia shozō uigurugo bunken no kenkyū-uigurumoji hyōki kanbun to uigurugo butten tekisuto* [Uighur Manuscripts in St. Petersburg Chinese texts in Uighur script and Buddhist Uighur texts] (Kyoto: Kyoto daigaku daigakuin bunka kenkyūjo, 2003).

35 Kasai, 'The Old Uyghur extract of the "Zhong Ahanjing"', pp. 77, 79.

If one compares those sentences with their Chinese original, it is evident that they correspond word for word to the Chinese ones, with the exception of the character *zhe* 者 at the end of the first sentence in the Chinese original; thus, they do not follow the Old Uyghur word order. To put the original word order into Old Uyghur, it would have to be changed in the following manner:

birök s(ä)n mintirdin äšidtiŋ ärsär: ulug ärän-lär-niŋ töz tüp bilgä bilig üzä bölmäk adırtlamak-ıg

birök s(ä)n mintirdin ulug ärän-lär-niŋ töz tüp bilgä bilig üzä bölmäk adırtlamak-ıg äšidtiŋ ärsär

otgurak bulup yeg üstünki kertgünčüg: mini-tä inčip öritgäy ärdiŋ ögrünčüg sävinčig tep:

yeg üstünki kertgünčüg otgurak bulup: mini-tä inčip ögrünčüg sävinčig öritgäy ärdiŋ tep:

Such Old Uyghur sentences in Chinese word order are also used in the Old Uyghur translation of *Abhidharmakośabhāṣyaṭīkātattvārtha-nāma*, a commentary on the *Abhidharmakośabhāṣya*, which is one of the most important Abhidharma texts. Toru Haneda, who scrutinized this Abhidharma commentary, pointed out that some important Old Uyghur word orders were still retained, although the use of Chinese word order is predominant in that text. He also assumed that the Uyghur Buddhists had referred to the original Chinese text when they used the texts written in the Chinese word order.[36]

This usage of an original Chinese text together with the corresponding Old Uyghur version can be supported further by the Old Uyghur Āgama texts, including the above-quoted *Madyamāgama*. The Āgama itself can be counted as one of the earliest types of Buddhist texts and therefore one of the fundamental Buddhist literature. Four kinds are known, which all consist of many various short sutras. Considering their characteristics, it is likely that they were all translated into Old Uyghur and, in fact, we have the Old Uyghur version of all four Āgamas. Curiously, most of the Old Uyghur Āgama texts which have been identified up to now are just extracts; no complete translations have yet been found. It is immediately apparent that these cannot be read like books, because they are cut up into fragmentary pieces which make them unintelligible without also having the complete text at hand. The sentences quoted above from the *Madyamāgama* are at least given as a whole, but there are some places where just a few words are translated or the Uyghur translation stops in the middle of a sentence. One clear example of this can be found in ll. 06–09 of the verso side of folio 1935.52.0005 preserved

36 Haneda, 回鶻訳本安慧の倶舎論実義疏 [Commentaire de l'Abhidarmakośa 倶舎論実義疏 par Sthiramati 安慧 en langue ouigoure], pp. 160–62.

in the Hedin Collection in Stockholm. Here some Chinese characters are also quoted before the corresponding Old Uyghur translation:

> 龍象 naga sudur: 烏陁姨 uday-i 東河 purvakosṭak ügüz-kä: 拭體著 ätözin sürtüp tonın käḍdi: 念作 sikadi tep
>
> > (Nāga-sūtra ... Udāyi ... To the Pūrvakoṣṭhakā-river ... He dried his body and put on his clothes. ... 'sikadi')[37]

If the reader only sees those sentences, it is obvious that the content of the text cannot be understood. But when one consults its Chinese counterpart, the whole context becomes clear:

> T. 26 *Zhong ahanjing*中阿含經, 608b02–09.
>
> 龍象經第二第三念誦
>
> 我聞如是。一時佛遊舍衛國在東園鹿子母堂。爾時世尊則於晡時從宴坐起。堂上來下告曰。烏陀夷。共汝往至東河澡浴。尊者烏陀夷白曰。唯然。於是世尊將尊者烏陀夷。往至東河脱衣岸上便入水浴。浴已還出拭體著衣。爾時波斯匿王有龍象名曰念作一切妓樂。
>
> > (Nāga-sūtra, the second. The third chanting.)
> >
> > (I heard the following. Once Buddha wandered in Śrāvastī and stayed in Pubbārāma Migāra-mātupāsāda. At that time, Tathāgata stood up from sitting meditation in the evening and came down from the [dharma]-hall and said: 'Udayi, I will go together with you to the Pūrvakoṣṭhakā river and bathe'. The worthy Udayi said: 'yes, please do so'. So Tathāgata led the worthy Udayi, went to the Pūrvakoṣṭhakā river, left their clothes on the shore, went into the water and bathed. After Udayi had finished bathing, he came out, dried his body and put on his clothes. At that time, Prasenajit had the elephant king whose name is *smṛti* and who could create all kinds of music.)[38]

Many such sentences or, rather, word quotations in Chinese can be found in the Old Uyghur Āgama texts and, like the sentence 拭體著 *ätözin sürtüp tonın käḍdi* above, the following Old Uyghur translation often goes further than the Chinese sentence quoted.[39] Thus, it is highly improbable that those texts could have easily been used without also referring to a complete text. Considering the fact that the Old Uyghur Āgamas mainly consist of collections of many short extracts, they were unlikely to have been read as books. Regarding the usage of them, Aydar

37 Kasai, 'The Old Uyghur extract of the "Zhong Ahanjing"', pp. 83–85.

38 The translation was made by me. The underlined phrases correspond to the Old Uyghur version.

39 The Old Uyghur translation corresponds to *shitizhuyi* 拭體著衣 '(He) rubbed his body and put on his clothes'. The Chinese quotation in the Old Uyghur text is incomplete.

Mirkamal made the noteworthy suggestion that those texts were used to learn how to translate from Chinese into Old Uyghur, and that Uyghur monks had used the Chinese texts during their learning as well.[40] In fact, some Āgama texts were written in cursive script on both sides of new blank paper, which shows that they were not produced as a formal scripture, although they are written carefully. Those characteristics lead us to the assumption that they could not have been a kind of a praxis notebook for someone's personal use, but perhaps were made for use inside a closed community, like a specific temple or Buddhist school.

Two Different Types of Abhidharma Translations in Old Uyghur

As mentioned above, Old Uyghur Āgamas and Abhidharmas were strongly influenced by Chinese Buddhist tradition — like the Āgamas, some Abhidharma texts even contain Chinese quotations. But there is another type of Abhidharma translation in which Brāhmī script is partly used, as well as Uyghur.[41] Brāhmī script stems from India; it was widespread in the Buddhist communities in Central Asia, and was used to write different languages. This script was probably introduced to the Uyghurs by the Tocharians, who used it for their own language.[42] In some Old Uyghur Buddhist texts, similarly to the way Chinese characters are used for Chinese words, Brāhmī script is employed only for chosen words — most of which are of Sanskrit origin.[43] In terms of the Old Uyghur Abhidharmas, both those containing Chinese quotations and those which are partly written in Brāhmī script have been found.[44] While the Abhidharma texts containing Chinese characters are well preserved, and some are carefully written in block script, those partly written in Brāhmī script are all fragmentary and written on the backside of the Chinese Buddhist texts in cursive script.[45] Those characteristics

40 Mirkamal, 回鶻僧人如何学习汉语 *Huihu sengren ruhe xuexi hanyu*, pp. 82–83. Further discussion with examples from the texts is shown in Kasai, Raschmann, and Zieme, 'Introduction', pp. 21–25.

41 The Old Uyghur fragments belonging to this text in the Berlin Turfan Collection have been gathered and edited, see Kasai, *Die altuigurischen Fragmente mit Brāhmī-Elementen*, Text Aa.

42 This topic has been well discussed; for an overview of notable contributions, see Kasai, *Die altuigurischen Fragmente mit Brāhmī-Elementen*, p. 7, fn. 2.

43 Kasai, 'Zum Gebrauch der Brāhmī-Schrift in alttürkischen buddhistischen Texten'.

44 The Abhidharma texts containing Chinese characters were mainly dealt with by M. Shōgaito, see Shōgaito, ウイグル文アビダルマ論書の文獻学的研究 *Uigurubun Abidaruma Ronsho No Bunkengakuteki Kenkyū*. Those which are partly written in Brāhmī script were gathered and edited by me, see Kasai, *Die altuigurischen Fragmente mit Brāhmī-Elementen*, Texte A. In addition, there are several Abhidharma texts which are completely written in Brāhmī script, see Maue, *Alttürkische Handschriften*, Part 19, pp. 165–85, no. 97–102.

45 A noteworthy example of a carefully written Abhidharma text would be the Old Uyghur version of the *Abhidharmakośabhāṣya*, preserved in the Museum of Ethnography in Stockholm, see Shōgaito, *The Uighur 'Abhidharmakośabhāṣya'*, facsimiles. About the latter type, see e.g. http://turfan.bbaw.de/dta/ch_u/images/chu7321versototal.jpg.

of texts partly written in Brāhmī script tend to indicate informal and personal use. But the *Abhidharmakośa-bhāṣya* — which is the best-preserved Abhidharma partly written in Brāhmī script — is in a booklet format,[46] suggesting that some specimens of this type were also produced for use over a longer time.

Because of the fragmentary condition of most texts in Brāhmī script, it is difficult to identify their counterpart in other languages. Only the *Abhidharmakośa-bhāṣya* mentioned above provides enough text for a comparative study. Examining it, one can conclude that its Old Uyghur version is most closely related to the Chinese one which was translated from Sanskrit by the famous Chinese monk Xuanzang 玄奘.[47] Xuanzang did not translate the Sanskrit original word by word into Chinese, but rather modified it in many points. Thus, the differences between the original Sanskrit and Xuanzang's Chinese version are easily visible and there is no doubt that the Old Uyghur version is closer to Xuanzang's. Nevertheless, the Old Uyghur version is not a direct translation of the Chinese version. In spite of making many changes, Xuanzang still kept the style of the Sanskrit original in so far as presenting first the doctrine of the Sarvāstivāda school in verse, followed by an explanation of it in prose. The Old Uyghur version, however, omits the important verse part completely.[48] Thus, anyone who read that text either needed to know the doctrinal part off by heart or have the complete text including the verse part at hand when using the Old Uyghur version. Considering its close relationship with the Chinese text, it is also possible that readers had access to the Chinese original in the latter case.

Therefore, it can be concluded that both types of Old Uyghur Abhidharmas were produced under the strong influence of Chinese texts, while the texts partly written in Brāhmī script suggest a close relationship to Tocharian and Indian cultures. In this context, it is important to mention that, up to now, no Old Uyghur Abhidharma translation has been found which contains both Brāhmī script and Chinese characters. This fact indicates that those two types of Abhidharmas were produced and used by two different groups. As mentioned above, over time several Buddhist schools were introduced to the Uyghurs. Of course, not all of them could gain a foothold in perpetuity. Many schools just had a passing popularity and later disappeared entirely, but a few of them consistently played a role among the Uyghur Buddhists. Tocharian Buddhism, which made a solid contribution towards establishing the Uyghurs' Buddhist worship, can surely be counted as one aspect of this.

---

46 See Kasai, *Die altuigurischen Fragmente mit Brāhmī-Elementen*, Text Aa. The text is now preserved in a fragmentary condition, but on the paper the trace of binding can be recognized. The reconstruction of the book format was shown by me, see Kasai, 'The Old Turkish Text Based on the "Abhidharmakośa-bhāsya"', p. 254, table II.

47 For the detailed discussion, see Kasai, 'The Old Turkish text based on the *Abhidharmakośa-bhāsya*', pp. 254–58.

48 The exact comparison between the Old Uyghur sentences and Chinese ones are shown in the edition, see Kasai, *Die altuigurischen Fragmente mit Brāhmī-Elementen*, Text Aa, pp. 32–55.

Considering that the preserved translations from Tocharian are from the pre-Mongol period, it appears that the Tocharians did not sustain the same level of influence as the Chinese. But the usage of Brāhmī script suggests that at least a small part of the Tocharian tradition probably remained in some Buddhist circles.

Most of the Old Uyghur translations that partly used Brāhmī script were written in cursive Uyghur script, which is one of the important characteristics for dating the copying of a work to the Mongol period. If we assume that the use of Brāhmī script began in the early period because of Tocharian influence, it must have had a second period of popularity later under Mongol rule. Given the characteristics of the texts in question, it is probable that the Chinese influence played a role in this second florescence of Brāhmī script. In fact, most of the Āgama texts containing Chinese characters were written in the same period, although the partial use of Chinese characters probably goes back further. Therefore, during the Mongol period the writing style in which Chinese characters were used as well as Uyghur script was popular among the Uyghur Buddhists who preferred the Chinese Buddhist tradition. The partial use of Brāhmī script was most likely due to the other group of Uyghur Buddhists being inspired by that writing style, although they themselves kept a part of Tocharian tradition alive.

Conclusion

This investigation of Old Uyghur translations has revealed that they contain an obvious Chinese Buddhist influence, which corresponds to the fact that most of the Old Uyghur Buddhist texts were translated from Chinese. Meanwhile, a part of the Tocharian tradition, which had been introduced in the early period, was still retained, even in the later period. This shows that different Buddhist groups coexisted in Uyghur Buddhist society. They seem to have had frequent contact with each other and been ready to accept each other. Flexibility and openness were also a characteristic of Uyghur society itself, located on the famous crossroads of Eurasia.

Bibliography

Primary Sources

Das Zusammentreffen mit Maitreya. Die ersten fünf Kapitel der Hami-Version der Maitrisimit, part II, ed. and trans. by Geng Shimin and Hans-Joachim Klimkeit, in cooperation with Helmut Eimer and Jens Peter Laut (Wiesbaden: Harrassowitz, 1988)

Maitrisimit. Faksimile der alttürkischen Version eines Werkes der buddhistischen Vaibhāṣika-Schule I, ed. by Annemarie von Gabain (Wiesbaden: Franz Steiner, 1957)

Secondary Studies

Abe, Takeo, 安部健夫, 西ウイグル国史の研究 *Nishi Uigurukokushi No Kenkyū [The History of the West Uyghur Kingdom]* (Kyoto: 中村印刷出版部, 1955)

Allsen, Thomas T., 'The Yüan Dynasty and the Uighurs of Turfan in the 13th Century', in *China Among Equals: The Middle Kingdom and Its Neighbors, 10th–14th Centuries*, ed. by Morris Rossabi (Berkeley: University of California Press, 1983), pp. 243–80

Brose, Michael C., *Subjects and Masters: Uyghurs in the Mongol Empire* (Bellingham: Center for East Asian Studies, Western Washington University, 2007)

Chavannes, Édouard, and Paul Pelliot, 'Un traité manichéen retrouvé en Chine', *Journal Asiatique*, 18 (1913), 99–199

Clark, Larry Vernon, 'The Conversion of Bügü Khan to Manichaeism', in *Studia Manichaica. IV. Internationaler Kongreß zum Manichäismus*, ed. by Ronald Eric Emmerick, Werner Sundermann, and Peter Zieme (Berlin: Akademie, 2000), pp. 83–123

Elverskog, Johan, *Uygur Buddhist Literature*, Silk Road Studies, 1 (Turnhout: Brepols, 1997)

Gabain, Annemarie von, 'Buddhistische Türkenmission', in *Asiatica: Festschrift Friedrich Weller: Zum 65. Geburtstag gewidmet von seinen Freunden, Kollegen und Schülern*, ed. by Johannes Schubert and Ulrich Schneider (Leipzig: Harrassowitz, 1954), pp. 161–73

Geng, Shimin, Jens Peter Laut, and Georges-Jean Pinault, 'Neue Ergebnisse der Maitrisimit-Forschung (I)', *Zeitschrift der Deutschen Morgenländischen Gesellschaft*, 154.2 (2004), 347–69

———, 'Neue Ergebnisse der Maitrisimit-Forschung (II): Struktur und Inhalt des 26. Kapitels', 内陸アジア言語の研究 *Nairiku Ajia Gengo No Kenkyū*, 19 (2004), 29–94

Golden, Peter Benjamin, *An Introduction to the History of the Turkic People: Ethnogenesis and State-Formation in Medieval and Early Modern Eurasia and the Middle East*, Turcologica, 9 (Wiesbaden: Harrassowitz, 1992)

Haneda 羽田亨, '回鶻訳本安慧の俱舎論実義疏', in 白鳥博士還暦記念東洋史論叢 *Shiratori Hakushi Kanreki Kinen Tōyōshi Ronsō* (Tokyo: 岩波書店, 1925)

———, 羽田博士史学論文集 下巻 言語·宗教篇 *Haneda Hakushi Shigaku Ronbunshū. Gekan Gengo, Shūkyō Hen* [French subtitle: Recueil des œuvres posthumes de Tōru Haneda. II: Études religieuses et linguistiques] (Kyoto: 東洋史研究会, 1958)

Ji, Xianling, Werner Winter, and Georges-Jean Pinault, *Fragments of the Tocharian A Maitreyasamiti-Nāṭaka of the Xinjiang Museum, China*, Trends in Linguistics, Studies and Monographs, 113 (Berlin: de Gruyter, 1998)

Kara, Georg, and Peter Zieme, *Fragmente tantrischer Werke in uigurischer Übersetzung*, Berliner Turfantexte, 7 (Berlin: Akademie, 1976)

———, *Die uigurischen Übersetzungen des Guruyogas "Tiefer Weg" von Sa-skya Paṇḍita und der Mañjuśrīnāmasaṃgīti*, Berliner Turfantexte, 8 (Berlin: Akademie, 1977)

Kara, György, 'L'ancien Ouigour dans le lexique Mongol', *Journal Asiatique*, 269.1/2 (1981), 317–23

Kasai, Yukiyo, 'Die uigurische Überlieferung der Legende von der Gründung des Tempels Baimasi', in *Aspects of Research into Central Asian Buddhism: In Memoriam Kōgi Kudara*, ed. by Peter Zieme, Silk Road Studies, 16 (Turnhout: Brepols, 2008), pp. 77–93

———, *Die uigurischen buddhistischen Kolophone*, Berliner Turfantexte, 26 (Turnhout: Brepols, 2008)

———, 'Der Einfluß der Vijñānavāda-Schule im alttürkischen Kommentar zum Vimalakīrti-Nirdeśa-Sūtra', in 突厥語文学研究–耿世民教授80華誕記念論文集 *Tujue Yuwenxue Yanjiu – Geng Shimin Jiaoshou 80 Huadan Jinian Lunwenji*, ed. by Dingjing Zhang and Abdurishid Yakup (Beijing: 中央民族大学出版社, 2009), pp. 298–316

———, *Die alttürkische Kommentar zum Vimalakīrtinirdeśa-Sūtra*, Berliner Turfantexte, 29 (Turnhout: Brepols, 2011)

———, 'The Old Turkish Text Based on the Abhidharmakośa-Bhāsya', in 西域–中亜語文学研究: 2012年中央民族大学主辨西域–中亜語文学国際学術研討会論文集 *Xiyu – Zhongya Yuwenxue Yanjiu. 2012 nian Zhongyang Minzu Daxue Zhuban Xiyu-Zhongya Yuwenxue Guoji Xueshu Yantaohui Lunwenji* [Studies in Central Asian Philology. Papers of the International Symposium on Central Asian Philology, November 2012, Beijing], ed. by Abdurishid Yakup阿不都熱西提·亜庫甫 (Shanghai: 上海古籍出版社, 2015), pp. 251–63

———, 'Zum Gebrauch der Brāhmī-Schrift in alttürkischen buddhistischen Texten', in *Kutadgu Nom Bitig. Festschrift für Jens Peter Laut zum 60. Geburtstag*, ed. by Elisabetta Ragagnin and Jens Wilkens (Wiesbaden: Harrassowitz, 2015), pp. 225–35

———, *Die altuigurischen Fragmente mit Brāhmī-Elementen*, Berliner Turfantexte, 38 (Turnhout: Brepols, 2017)

———, 'The Old Uyghur Extract of the Zhong ahanjing (T. 26): Three of Eight Folios Preserved in the Hedin Collection (1935.52.0004–0006)', in *The Old Uyghur Āgama Fragments Preserved in the Sven Hedin Collection, Stockholm*, ed. by Yukiyo Kasai, Simone-Christiane Raschmann, Håkan Wahlquist, and Peter Zieme, Silk Road Studies, 15 (Turnhout: Brepols, 2017), pp. 73–112

Kasai, Yukiyo, Simone-Christiane Raschmann, and Peter Zieme, 'Introduction', in *The Old Uyghur Āgama Fragments Preserved in the Sven Hedin Collection, Stockholm*, ed. by Yukiyo Kasai, Simone-Christiane Raschmann, Håkan Wahlquist, and Peter Zieme, Silk Road Studies, 15 (Turnhout: Brepols, 2017), pp. 17–36

Kitsudō, Kōichi, 橘堂晃, 'ウイグル語訳『観弥勒上生兜率天経賛』について [On the Uyghur Version Guan Mile Shangsheng Doushuaitian Jing Zan]', *Bukkyō Shigaku Kenkyū* 佛教史學研究, 51.1 (2008), 24–46

———, 'Liao Influence on Uigur Buddhism', in *Studies in Chinese Manuscripts: From the Warring States Period to the 20th Century*, ed. by Imre Galambos (Budapest: Eotvos Lorans University. Institute of East Asian Studies, 2013), pp. 225–47

———, 'ウイグル文慈恩宗文献「大唐三藏行跡讚」[Uyghur Yogacārā Text Datang Sanzang Xingji Zan]', 敦煌寫本研究年報 *Tonkōshahon Kenkyū Nenpō*, 10 (2016), 371–86

Kudara, Kōgi 百済康義, '妙法蓮華経玄賛のウイグル訳断片 [Old Uyghur Fragment of the Miao-Fa-Lian-Hua-Jing Xuan-Zan]', in 内陸アジア·西アジアの社会と文化 *Nairiku Ajia, Nishi Ajia No Shakai to Bunka [Society and Culture of Inner Asia and the Muslim World]*, ed. by 護雅夫 Masao Mori (Tokyo: 山川出版社, 1983), pp. 185–207

Laut, Jens Peter, *Der frühe türkische Buddhismus und seine literarischen Denkmäler* (Wiesbaden: Harrassowitz, 1986)

Laut, Jens Peter, and Jens Wilkens, *Alttürkische Handschriften, Teil III. Die Handschriftenfragmente der Maitrisimit aus Sängim und Murtuk in der Berliner Turfansammlung* (Stuttgart: Franz Steiner, 2017)

Mackerras, Collin, *The Uighurs* (Cambridge: Cambridge University Press, 1990)

Maue, Dieter, *Alttürkische Handschriften. Teil 19. Dokumente in Brāhmī und tibetisher Schrift Teil 2* (Stuttgart: Franz Steiner, 2015)

Maue, Dieter, and Klaus Röhrborn, 'Ein "buddhistischer Katechismus" in alttürkischer Sprache und tibetischer Schrift (Teil I)', *Zeitschrift der Deutschen Morgenländischen Gesellschaft*, 134.2 (1984), 286–313

———, 'Ein "buddhistischer Katechismus" in alttürkischer Sprache und tibetischer Schrift (Teil II)', *Zeitschrift der Deutschen Morgenländischen Gesellschaft*, 135.1 (1985), 68–91

Mirkamal, Aydar, '回鶻僧人如何学习汉语 *Huihu sengren ruhe xuexi hanyu?*' [How Did Uyghur Monks Learn Chinese?], *Xinjiang Shifan Daxue Xuebao* 新疆师范大学学报, 35.1 (2014), 81–86

Mironov, Nikolaj D., 'Kuchean Studies', *Rocznik Orientalistyczny*, 6 (1929), 89–169

Moerloose, Eddy, 'Sanskrit Loan Words in Uighur', *Journal of Turkish Studies*, 4 (1980), 61–78

Moriyasu, Takao, 森安孝夫, 'ウィグルの西遷について [French subtitle: Nouvel examen de la migration des ouïgours au milieu du IXe siècle]', *Tōyō Gakuhō* 東洋学報, 59.1/2 (1977), 105–30

———, 'チベット文字で書かれたウィグル文仏教教理問答 (P. t. 1292) の研究 [French subtitle: 'Études sur un catéchisme bouddhique ouigour en écriture tibétaine (Pt. 1292)]', *Ōsaka Daigaku Bungakubu Kiyō* 大阪大学文学部紀要, 25 (1985), 1–85

———, *Die Geschichte des uigurischen Manichäismus an der Seidenstraße. Forschungen zu manichäischen Quellen und ihrem geschichtlichen Hintergrund* (Wiesbaden: Harrassowitz, 2004)

Oda, Juten, 小田壽典, 'トルコ語本八陽経写本の系譜と宗教思想的問題 [English subtitle: Some Problems on Manuscripts of the Eight Lights Sūtra in Uygur]', 東方学 *Tōhōgaku*, 55 (1978), 118–04

———, 'A Fragment of the Uighur *Avalokiteśvara-Sūtra* with Notes', in *Turfan, Khotan und Dunhuang. Vorträge der Tagung 'Annemarie v. Gabain und die Turfanforschung', veranstaltet von der Berlin-Brandenburgischen Akademie der*

Wissenschaften in Berlin, ed. by Ronald Eric Emmerick, Werner Sundermann, Ingrid Warnke, and Peter Zieme (Berlin: Aakademie, 1996), pp. 229–43

———, 'On Manichaean Expressions in the Säkiz Yükmäk Yaruq', in *Splitter aus der Gegend von Turfan. Festschrift für Peter Zieme anläßlich seines 60. Geburtstags* (Istanbul: Mehmet Ölmez and Simone-Christiane Raschmann, 2002), pp. 179–98

———, *A Study of the Buddhist Sūtra called Säkiz Yükmäk Yaruq or Säkiz Törlügin Yarumiš Yaltrïmiš in Old Turkic*, Berliner Turfantexte, 33 (Turnhout: Brepols, 2015)

Röhrborn, Klaus, 'Die alttürkische "Maitrisimit" — Textbuch für theatralische Darstellungen?', in *Memoriae Munusculum. Gedenkband für Annemarie v. Gabain* (Wiesbaden: Harrassowitz, 1994), pp. 99–103

———, 'Die alttürkische Xuanzang-Vita: Biographie oder Hagiographie?', in *Bauddhavidyāsudhākaraḥ. Studies in Honour of Heinz Bechert on the Occasion of His 65th Birthday*, ed. by Petra Kieffer-Pülz and Jens-Uwe Hartmann, Indica et Tibetica (Swisttal-Odendorf: Indica et Tibetica, 1997), pp. 551–57

———, 'Zum manichäischen Einfluß im alttürkischen Buddhismus', in *Studia Manichaica. IV. Internationaler Kongreß zum Manichäismus, Berlin, 14.-18. Juli 1997*, ed. by Ronald Eric Emmerick, Werner Sundermann, and Peter Zieme (Berlin: Akademie, 2000), pp. 494–99

Rong, Xinjiang, 'The Relationship of Dunhuang with the Uighur Kingdom in Turfan in the Tenth Century', in *De Dunhuang à Istanbul. Hommage à James Russell Hamilton*, ed. by Louis Bazin and Peter Zieme, Silk Road Studies, 5 (Turnhout: Brepols, 2001), pp. 275–98

Saguchi, Tōru, 佐口透, 'モンゴル人支配時代のウイグリスタン (上) (下) [Uiguristan under the Mongol Rule (I, II)]', *Shigaku Zasshi* 史学雑誌, 54.8/9 (1943), 1–71, and 72–97

Shōgaito, Masahiro, 庄垣内正弘, '古代ウイグル語'におけるインド来源借用語彙の導入経路について [English subtitle: On the Routes of the Loan Words of Indic Origin in the Old Uigur Language], アジア・アフリカ言語文化研究 *Ajia, Afurika Gengo Bunka Kenkyū*, 15 (1978), 79–110

———, 'Uighur Influence on Indian Words in Mongolian Buddhist Texts', in *Indien und Zentralasien. Sprach- und Kulturkontakt. Vorträge des Göttinger Symposions vom 7. bis 10. Mai 2001*, ed. by Sven Bretfeld and Jens Wilkens (Wiesbaden: Harrassowitz, 2003), pp. 119–43

———, 庄垣内正弘, ロシア所蔵ウイグル語文献の研究ーウイグル文字表記漢文とウイグル語仏典テキストー *Roshia shozō uigurugo bunken no kenkyū-uigurumoji hyōki kanbun to uigurugo butten tekisuto* [Uighur Manuscripts in St. Petersburg Chinese texts in Uighur script and Buddhist Uighur texts] (Kyoto: Kyoto daigaku daigakuin bunka kenkyūjo, 2003)

———, ウイグル文アビダルマ論書の文獻学的研究 *Uigurubun Abidaruma Ronsho No Bunkengakuteki Kenkyū* [Uighur Abhidharma Texts: A Philological Study] (Kyoto: 松香堂出版, 2008)

———, *The Uighur 'Abhidharmakośabhāṣya' Preserved at the Museum of Ethnography in Stockholm*, Turcologica (Wiesbaden: Harrassowitz, 2014)

Sinor, Denis, Geng Shimin, and Yevgenij Ivanovič Kychanov, 'The Uighurs, the Kyrgyz and the Tangut (Eighth to the Thirteenth Century)', in *History of Civilizations of Central Asia*, vol. IV: *The Age of Achievement: A.D. 750 to the End of the Fifteenth Century*, part I: *The Historical, Social and Economic Setting*, ed. by Muchamed Sajfitdinovič Asimov and Clifford Edmund Bosworth (Paris: UNESCO, 1998), pp. 191–214

Tardieu, Michel, *Manichaeism*, trans. by M. B. DeBevoise (Urbana: University of Illinois Press, 2008)

Tremblay, Xavier, 'The Spread of Buddhism in Serindia-Buddhism Among Iranians, Tocharians and Turks Before the 13th Century', in *The Spread of Buddhism*, ed. by Ann Heirman and Stephan Peter Bumbacher (Leiden: Brill, 2012), pp. 75–129

Ueyama, Daishun, 上山大俊, '敦煌と曇曠の仏教学 [Dunhuang and Tankuang's Buddhism]', 東方学報 京都 *Tōhō Gakuhō Kyoto*, 35 (西明寺学僧曇曠と敦煌の仏教学 [Monk Tankuang in the monastry Xi ming and Buddhism in Dunhuang]) (1964), 17–83

———, '大蕃国大徳三蔵法師沙門法成の研究' [Study on Dafanguo Dade Sanzang Fashi, Facheng], 東方学報 京都 *Tōhō Gakuhō Kyoto*, 38 (大蕃国大徳三蔵法師法成の人と業績 [Dafanguo dade sanzang fashi Facheng and his achievements]) (1990), 84–246

———, 敦煌仏教の研究 *Tonkō Bukkyō No Kenkyū* [Study of Buddhism in Dunhuang] (Kyoto: Hōzōkan, 1990)

Wilkens, Jens, 'Hatten die alten Uiguren einen buddhistischen Kanon?', in *Kanonisierung und Kanonbildung in der Asiatischen Religionsgeschichte*, ed. by Max Deeg, Oliver Freiberger, and Christoph Kleine (Wien: Verlag der Österreichischen Akademie der Wissenschaften, 2011), pp. 345–78

———, 'Der "Neutag" und die "Maitrisimit" — Probleme der zentralasiatischen Religionsgeschichte', in *Die Erforschung des Tocharischen und die alttürkische Maitrisimit. Symposium anlässlich des 100. Jahrestages der Entzifferung des Tocharischen, Berlin, 3. und 4. April 2008*, ed. by Yukiyo Kasai, Abdurishid Yakup, and Desmond Durkin-Meisterernst, Silk Road Studies, 17 (Turnhout: Brepols, 2013), pp. 375–401

———, *Buddhistische Erzählungen aus dem alten Zentralasien. Edition der altuigurischen Daśakarmapathāvadānamālā*, Berliner Turfantexte, 37 (Turnhout: Brepols, 2016)

Yoshida, Yutaka, 'Die buddhistischen sogdischen Texte in der Berliner Turfansammlung und die Herkunft des buddhistischen sogdischen Wortes für Bodhisattva. Zum Gedenken an Prof. Kōgi Kudaras Arbeiten an den sogdischen Texten', trans. by Yukiyo Kasai and Christiane Reck, *Acta Orientalia Academiae Scientiarum Hungaricae*, 61.3 (2008), 325–58

Yüsüp, Israpil 玉素甫·伊斯拉菲爾, Dolqun Qämbiri 多魯坤·闞白爾, and Abduqeyum Xoğa 阿不都克由木·霍加, 回鶻文弥勒会見記 *Huihuwen Mile Huijianji [Maitrisimit in Old Uyghur]* (Urumqi: Shinjang Khălq Năshriyati 新疆人民出版社, 1988)

Zieme, Peter, and György Kara, *Ein uigurisches Totenbuch. Nāropas Lehre in uigurischer Übersetzung von vier tibetischen Traktaten nach der Sammelhandschrift aus Dunhuang British Museum Or. 8212(109)*, ed. by Peter Zieme and György Kara, Bibliotheca Orientalis Hungarica, 22 (Budapest: Akadémiai Kiadó, 1979)

JENS HØYRUP

Advanced Arithmetic from Twelfth-Century al-Andalus, Surviving Only (and Anonymously) in Latin Translation?

A Narrative that Was Never Told

In memory of *Michel Olsen (1934–2013)*
and *Gunver Kelstrup (1935–2015)*
dear friends

Background

The writings of ibn Rušd, as anybody even slightly familiar with Latin or Hebrew medieval philosophy knows, had a major influence on both traditions. His impact on Arabic thought, on the other hand, was modest — more modest, indeed, than can be explained from his date or from al-Ghazālī's attack on the incoherence of philosophers (to which he wrote an answer). Similarly, while al-Muʾtaman's eleventh-century *Kitāb al-Istikmāl* gave rise to further work by Arabic mathematicians,[1] Jābir ibn Aflaḥ's twelfth-century work in astronomy and spherics is much better known from Hebrew and Latin translations than in Arabic.[2]

The obvious explanation for this is that al-Andalus was already cut off from the corresponding courses of Islamic scholarly life. Not all courses, as we know — ibn al-Yāsamīn's work, partly done in al-Andalus and partly in the Maghreb, did survive in Arabic (while leaving no clearly recognizable traces in Latin or in Romance vernaculars); but this work was integrated with an interest that thrived in the vicinity of *madrasah* learning as established in the Maghreb — in the terminology of A. I. Sabra, it represented 'naturalized' knowledge.[3] It is therefore tempting to ask

1 Djebbar, 'Deux mathématiciens peu connus de l'Espagne du XIᵉ siècle', p. 82.
2 Lorch, 'Jābir ibn Aflaḥ', p. 39.
3 Sabra, 'The Appropriation and Subsequent Naturalization of Greek Science'.

Jens Høyrup • (jensh@ruc.dk) is a professor emeritus, Section for Philosophy and Science Studies, Roskilde University; his main research areas are the history of Mesopotamian and European medieval mathematics and pre-Modern practitioners' mathematics.

Premodern Translation: Comparative Approaches to Cross-Cultural Transformations, ed. by Sonja Brentjes and Alexander Fidora, CAT 2 (Turnhout: Brepols, 2021), pp. 33–61

10.1484/M.CAT-EB.5.122133

whether a fate similar to that of ibn Rušd's philosophy befell some further branches of al-Andalus's learning.

One such branch might be theoretical arithmetic. As Ahmed Djebbar observed,[4] there was:

> in Spain and before the eleventh century, a solid research tradition in arithmetic, whose starting point seems to have been the translation made by Thābit ibn Qurra of Nicomachos's *Introduction to Arithmetic*.[5]

Arabic sources do not provide any evidence that this tradition lived on into the twelfth century. However, as it has dawned upon me over the last decade, three different Latin and one Tuscan vernacular sources which I had worked on without at first seeing any connection between them may be witnesses not only of survival, but of impressive expansion. Since not all of what I have written on these occasions can be expressed better (at least not by me, at this moment), what follows will contain a certain amount of auto-plagiarism.

The Unknown Heritage — The Simple Version

My first example is a theoretical elaboration of the solution to a stunning recreational problem — the 'unknown heritage'.[6] The standard version of this problem runs as follows: a father leaves to his first son 1 monetary unit and $\frac{1}{n}$ (*n* usually being 7 or 10) of what remains; to the second he leaves 2 units and $\frac{1}{n}$ of what remains, and so on. In the end all his sons receive the same amount, and nothing remains. The solution is that there are $n-1$ sons, each of whom receives $n-1$ monetary units. Alternatively, the fraction is given first and the arithmetically increasing amount afterwards, in which case $n-1$ sons are given *n* monetary units each.

Our earliest source for both variants of this simple version of the problem is Chapter 12 of Fibonacci's *Liber abbaci*[7] — 'simple' only in comparison with the 'sophisticated' version to which we shall return, not in comparison with the other recreational problems dealt with in that same chapter. It is possible to find the only *possible* solution by algebra or by a double false position from the equality of the first two shares, but in order to show that this really *is* a solution one has to perform a stepwise control. It is also feasible to deduce the only possible solution by elementary means from the equality of the last two shares, but this solution suffers from the same defect as the previous one; moreover, it appears to have escaped all medieval

4 My translation, as all translations in the following with no identified translator.

5 Djebbar, 'Deux mathématiciens peu connus de l'Espagne du XIe siècle', p. 86.

6 Høyrup, 'The "Unknown Heritage"'.

7 Fibonacci, *Il Liber abbaci*, ed. by Boncompagni, p. 279. Since all the copious references to the *Liber abbaci* in the following refer to this edition, I shall only indicate them by page number.

and early modern authors presenting the problem, and even all modern historians who have worked on it.[8]

This 'simple' problem is frequently found in Italian abbacus books from the early fourteenth century onward, mostly the first variant but also sometimes the second, and the corresponding semi-simple variants where the absolute contributions are not i ($i = 1, 2, \ldots$) but $i\varepsilon$ ($i = n, n+1, \ldots$), which corresponds to taking ε rather than 1 as the monetary unit and skipping the first $n-1$ heirs. Such semi-simple variants are also presented by Fibonacci. Most of the abbacus authors merely state the solutions, but Jacopo da Firenze offers a full numerical check in his *Tractatus algorismi,* written in Montpellier in 1307,[9] that the solution is correct (while speaking about picking oranges from a garden instead of an inheritance). The *Istratti di ragioni*[10] – a problem collection from *c.* 1440 which claims to go back to Paolo dell'Abbaco (*c.* 1340) and at least likely to copy material from that period — finds the possible solution by means of a double false position applied to the equality of the first two shares.[11]

Of particular interest is, on the one hand, the appearance of the problem in Maximos Planudes's late thirteenth-century *Calculus according to the Indians, Called the Great;*[12] and, on the other, its apparent absence from Arabic sources (even though two of them contain a derived and simplified version of it).

Planudes presents us with the second occurrence we know about, preceded only by that in the *Liber abbaci.* It follows after an exposition of how to calculate using Hindu-Arabic numerals, just before a discussion of how to 'find a figure equal in perimeter to another figure and a multiple of it in area' — that is, for a given n to find two rectangles[13] $\sqsubset\sqsupset(a,b)$ and $\sqsubset\sqsupset(c,d)$ such that $a+b = c+d$, $n \cdot ab = cd$ (a, b, c and d being tacitly assumed to be integers). It serves as an illustration of this observation or theorem:[14]

8 Since the last son (say, no. n) leaves nothing, the remainder r_n of which he takes the fraction $^1/_d$ must be 0 (if not, $(1-{^1/_d})r_n$ would be left over). But since each visitor gets as many units as his number before taking $^1/_d$ of the remainder, the nth son gets n units, and so therefore do all the others. But the second-last visitor (no. $n-1$) only gets $n-1$ units before taking the fraction $^1/_d$ of the remainder r_{n-1}. Therefore, this fraction must be 1 (he has already got $n-1$ but should have n). Further, he leaves n to the last visitor. In consequence r_{n-1} is $n+1$. $^1/_d$ of this number being 1 ($^{n+1}/_d = 1$), n must be $d-1$. It is hard to believe that Euler should have overlooked this in his *Élémens d'algebre*, pp. 488–91; but if he has seen it, he does not tell his readers. After all, he is teaching algebra, and may have seen no reason to divulge that algebra is not needed.

9 Jacopo da Firenze, *Tractatus Algorismi*, ed. and trans. by Høyrup, pp. 360–61; for my arguments that the Vatican manuscript Vat. lat. 4826 is most likely a copy of Jacopo's original treatise, whereas the two other surviving manuscripts descend from a revised version, see pp. 6–25.

10 Paolo Dell'Abaco, *Trattato d'aritmetica*, ed. by Arrighi, pp. 140–41.

11 A number of other abbacus occurrences are listed in Høyrup, 'The "Unknown Heritage"', pp. 628–30, 640–41 (after 2008 I have noticed quite a few more, none of them offering anything new).

12 Maximos Planudes, *Le Grand calcul selon les Indiens*, ed. and trans. by Allard, pp. 191–94.

13 Χωρίον, here translated 'figure', may actually have the more restricted meaning 'rectangular area'.

The statement of the problem and the first of two solutions are found almost verbatim in the pseudo-Heronic *Geometrica*, Chapter 24, see Heron of Alexandria, *Definitiones cum variis collectionibus. Heronis quae feruntur Geometrica*, ed. and trans. by Heiberg, pp. 414–17.

14 I here attempt a very literal translation from the Greek text.

> When a unit is taken away from any square number, the remainder is measured by two numbers multiplied by each other, one smaller than the side of the square by a unit, the other larger than the same side by a unit. As for instance, if from 36 a unit is taken away, 35 is left. This is measured by 5 and 7, since the quintuple of 7 is 35. If again from 35 I take away the part of the larger number, that is the seventh, which is then 5 units, and yet 2 units, the remainder, which is then 28, is measured again by two numbers, one smaller than the said side by two units, the other larger by a unit, since the quadruple of 7 is 28. If again from the 28 I take away 3 units and its seventh, which is then 4, the remainder, which is then 21, is measured by the number which is three units less than the side and by the one which is larger by a unit, since the triple of 7 is 21. And always in this way.

This description does not refer explicitly to pebbles or other counters, but it points rather unambiguously at something like Figure 1 (for simplicity showing a 5 × 5 square); in fact, this is the kind of diagram I made spontaneously when first encountering this problem and the stepwise calculation in Jacopo's *Tractatus*. Even straightforward application of algebraic symbols does not easily show that the procedure goes on in such a way that exactly nothing remains in the end.[15]

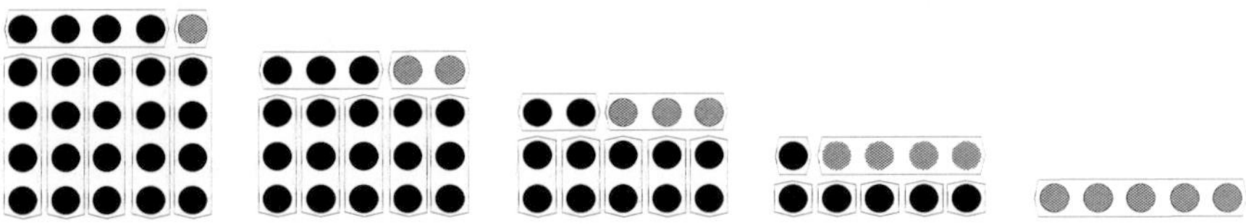

Figure 1

The pebble pattern not only allows us to understand the solution, it is also a likely basis for the discovery that the counter-intuitive problem was *possible*. Since pebble arguments were current in Ancient Greek mathematics and in view of the vicinity to material of Ancient or at most of early Byzantine date it is highly plausible that the problem originated in late Greek Antiquity or in a Byzantine context.[16]

The quasi-occurrences in Arabic sources[17] show beyond reasonable doubt that the problem was not transmitted to Fibonacci through mainstream Arabic recreational mathematics. One of them is in ibn al-Yāsamīn's *Talqīḥ al-afkār fī'l ʿamali bi rušūm al-ghubār* ('Fecundation of thoughts through use of *ghubār* numerals'), written in Marrakesh in *c.* 1190, that is, before the first version of the *Liber abbaci* from 1202. It runs as follows:

15 The proof will have to build on a chain of identities $n\cdot(n-p+1) = p+(n+1)\cdot(n-p)$, p increasing from 1 to n.

16 I have observed no traces of the problem or of the underlying theorem in such ancient sources as normally offer veiled references to mathematical practitioners' knowledge or problems (those Platonizing or Pythagoreanizing writers who tried to transform the mathematics they understood into 'wisdom', cf. Høyrup, 'Which Kind of Mathematics Was Known and Referred To'). It is therefore unlikely that the problem was (widely) known before, say, 200 CE.

17 I thank Mahdi Abdeljaouad for tracing them and supplying me with French translations, which I here translate into English.

> An inheritance of an unknown amount. A man has died and has left at his death to his children an unknown amount. He has left to one of the children one dinar and the seventh of what remains, to the second child two dinars and the seventh of what remains, to the third three dinars and the seventh of what remains, to the fourth child 4 dinars and the seventh of what remains, to the fifth child 5 dinars and the seventh of what remains, and to the sixth child what remains. He has ordered that the shares must be identical. What is the sum?
> The solution is to multiply the number of children by itself, giving you 36 as the unknown sum. This is a rule that recurs in all problems of the same type.

The other version comes from the *al-Maʿūna fī ʿilm al-ḥisāb al-hawāʾī* ('Assistance in the science of mental calculation') written by ibn al-Hāʾim (1352–1412, Cairo, Mecca and Jerusalem, and familiar with ibn al-Yāsamīn's work):

> An amount of money has been diminished by one dirham, and the seventh [of what remains]; by two dirhams, and then the seventh of what remains; then three dirhams and the seventh of what remains; then four dirhams and the seventh of what remains; then five dirhams and the seventh of what remains. In the end, six remain.
> Solution: take the square of the six that remain, to find the amount which was asked for.

Ibn al-Yāsamīn, we see, does not tell his readers that the last share is determined according to the same rule as the preceding ones, and ibn al-Hāʾim does state that the shares are equal. In fact, both pieces of information are indeed superfluous. The number of shares is given in both versions, and both are 'Chinese box problems' that can be solved by backward calculation; none the less, both still use the same rule as Fibonacci's and Planudes's version of the simple problem.

Similar backward calculations could be made for fractions that change and for absolutely defined contributions that are not in an arithmetical progression. However, the rule is only valid for a constant fraction $\frac{1}{N+1}$, where N is the given number of shares, and if the absolutely defined contributions are $1+(i-1)$. We are allowed to conclude that the Arabic problem descends from the 'Christian' problem, and that it is the outcome of an unsuccessful attempt to assimilate it to a more familiar structure.

Mathematicians from the Maghreb or al-Andalus[18] had thus come to know about the problem type before the *Liber abbaci* was thought of; but their reference to a rule that is adapted to the 'Christian' version shows that this latter version, with its unknown value of N, was not derived from the box-problem versions known in Arabic. Fibonacci and Planudes show us the original, which must nonetheless antedate ibn al-Yāsamīn and, therefore, both of them.

---

18 Ibn al-Yāsamīn's 'all problems of the same type' seems to prove that he was not alone in his area to know about them. Since he had been active in Morocco as well as al-Andalus, this area could be either, or both.

The Sophisticated Version of the Unknown Heritage

Fibonacci, however, presents us with more, and that is where things become interesting. A notation will be handy for our further discussion. The division of a number (or an abstract amount of *dragmae* which, in the language inherited from Arabic algebra, amounts to the same as Diophantos's *monades*) in such a way that each of the successive parts receives first $\alpha+i\varepsilon$, $i = 0, 1, 2, \ldots$, and afterwards φ times what remains at hand ($\varphi<1$, but not necessarily an aliquot part $\frac{1}{n}$) we shall designate $(\alpha,\varepsilon|\varphi)$. A division where instead φ times what is available is taken first, and afterwards an absolutely defined amount $\alpha+i\varepsilon$, we shall designate $(\varphi|\alpha,\varepsilon)$. In this notation, Fibonacci offers the following problems:

$(1,1\|\frac{1}{7})$	$(1,1\|\frac{2}{11})$	$(2,3\|\frac{6}{31})$	$(3,2\|\frac{5}{19})$
$(\frac{1}{7}\|1,1)$	$(4,4\|\frac{4}{11})$	$(\frac{6}{31}\|2,3)$	$(\frac{5}{19}\|3,2)$
$(3,3\|\frac{1}{7})$	$(\frac{2}{11}\|1,1)$		
$(\frac{1}{7}\|3,3)$	$(\frac{2}{11}\|4,4)$		

The column on the left contains the two variants of the simple version, together with the equally simple variant where the monetary unit is 3 *bizantii* instead of 1. Here, everything is stated in terms of a father distributing his possessions to his sons.

The remaining columns speak about dividing a number or a number of *dragmae* in the ways indicated. Here, all the shares are similarly stated to be equal, but with the difference that the last share may be fractional. The problems in the second column are dealt with according to the rule for the simple version, with the unexplained trick that $\frac{2}{11}$ is understood as $\frac{1}{5\frac{1}{2}}$. Then the number of shares turn out to be 4½, meaning that the last (half)share is only half of the others.

This trick, however, does not work in the third and the fourth column. For the problem $(2,3|\frac{6}{31})$ Fibonacci finds the only possible solution by means of *regula recta*, that is, rhetorical first-degree algebra: he calls the number to be divided a *thing*, computes the first two shares and equates them (p. 280). The number to be divided is then found (if expressed in a formula that follows the calculations step by step) to be:

$$T = \frac{q^2(\alpha+\epsilon) - (q-p)q\alpha - (q-p)p\alpha - (\alpha+\varepsilon)pq}{p^2}. \tag{1}$$

Afterwards Fibonacci makes a step-by-step control calculation, showing that all the shares are indeed equal. In the end he claims to extract this rule from the calculation (*ex hac quidem investigatione talem extraxi regulam*):[19]

$$T = \frac{[(\varepsilon-\alpha)q + (q-p)\alpha]\cdot(q-p)}{p^2}, \tag{2a}$$

$$N = \frac{(\epsilon-\alpha)q + (q-p)\alpha}{\epsilon p}, \tag{2b}$$

$$\Delta = \frac{\epsilon(q-p)}{p}, \tag{2c}$$

19 Fibonacci expresses himself in terms of the numerical values belonging with his paradigmatic example; but he identifies these so precisely that the translation into symbols in unambiguous.

N being the number of shares and Δ the value of each of them.

This is a case of mild fraud. With the techniques at Fibonacci's disposal (1) might be transformed into

$$T = \frac{[q(\alpha + \epsilon) - (p - q)\alpha] \cdot (q - p)}{p^2} \qquad (1^*)$$

and possibly into

$$T = \frac{[\varepsilon q - \alpha p] \cdot (q - p)}{p^2} \qquad (1^{**})$$

But never into (2a), however much *we*, with our more convenient tools, can prove their algebraic identity. We may conclude that Fibonacci took over a rule whose derivation he did not know, and claimed it to be a consequence of his own solution.

This is confirmed by the solution he offers (this time without calculations of his own) for the case of $(3{,}2|\,\frac{5}{19})$. Formula (1) would apply smoothly to this case, but formula (2a), which Fibonacci claims to have derived, contains a factor $\varepsilon - \alpha$, which is impossible if you have no developed conceptualization of negative numbers; it is not sufficient to know in practice how to make simple additive operations with them, as Fibonacci did. Instead he therefore gives the following rule:

$$T = \frac{[(q - p)\alpha - (\alpha - \varepsilon)] \cdot (q - p)}{p^2} \qquad (3)$$

which subtracts $(\alpha - \varepsilon)q$ instead of adding $(\varepsilon - \alpha)q$.

For the case $(\frac{6}{31}|2{,}3)$ Fibonacci gives the solution

$$T = \frac{[(\varepsilon -)q + (q - p)]}{p^2} \qquad (4)$$

and for $(\frac{5}{19}\,|3{,}2)$

$$T = \frac{[(q - p)\alpha - (\alpha - \varepsilon)q] \cdot q}{p^2} \qquad (5)$$

also unconnected to his algebraic calculation.

I know of no surviving Italian text from the fourteenth or the early decades of the fifteenth century where these sophisticated versions of the problem turn up. The first trace after Fibonacci (but a mere trace) is Cardano's treatment of a problem $(\frac{1}{7}|100{,}100)$ in the *Practica arithmetice et mensurandi singularis*.[20] Instead of the usual rule, he gives the solution

$$T = \frac{[(q - p)q] \cdot \alpha}{p^2} \qquad (6a)$$

20 Cardano, *Practica arithmetice, et mensurandi singularis*, fol. FF ii^r.

which follows from Fibonacci's rules (4) or (5) if we put $\varepsilon = \alpha$ and invert two factors (an inversion which could not be made in the full rules (4) or (5) if $\alpha \neq \varepsilon$). His rule for the number of heirs is

$$N = q - p\,. \tag{6b}$$

Since in (6a) Cardano dutifully finds 1·1 (p^2) and divides by it, (6b) must be considered a mistake for

$$N = \frac{q-p}{p}\,. \tag{6b*}$$

Both this mistake and the inversion of factors in (6a) seem to rule out the notion that Cardano built directly on the *Liber abbaci*. In spite of our ignorance, it seems likely that there was some kind of circulation of the sophisticated versions.

This is confirmed by Barthélemy de Romans's *Compendy de la praticque des nombres*, a Franco-Provençal treatise from *c.* 1467,[21] in which these *progressions composées* (as Barthélemy calls them) constitute the high point.[22] Some of the φ-values known from the *Liber abbaci* are repeated by Barthélemy, which has led Spiesser to find it 'very plausible' that Fibonacci was the direct source.[23] A closer analysis shows, however, that the coincidences are very far from being statistically significant;[24] moreover, when Barthélemy repeats a φ-value known from Fibonacci, the appurtenant α and ε are different in 8 out of 9 cases; finally, Barthélemy's global approach as well as his rules are quite different from what we know from Fibonacci, even in that part of the text where Barthélemy seems to borrow from precursors.[25]

If Fibonacci's own text shows him to have borrowed (which he does not deny in general, even though here he pretends to have extracted the rules from his own calculation), and if both Cardano's hearsay knowledge and Barthélemy's real acquaintance with the sophisticated version of our problem do not depend on the *Liber abbaci* — then where does it come from?

Fibonacci himself tells the reader (p. 1) that he had studied abbacus matters, first (as a boy) in Bejaïa, and afterwards in 'Egypt, Syria, Greece [i.e., Byzantium], Sicily and Provence'. Above I argued that the simple version is likely to have originated in

---

21 We only possess a revised redaction from 1476 due to Mathieu Préhoude, which has been edited with a modern French translation by Maryvonne Spiesser, see Barthélemy de Romans, *Une arithmétique commerciale du xve siècle*, ed. and trans. by Spiesser. The original treatise was said by Barthélemy to be written in order to provide clearer understanding to those who had read an earlier work of his (p. 225).

22 The pertinent part of the text is found in Barthélemy de Romans, *Une arithmétique commerciale du xve siècle*, ed. and trans. by Spiesser, pp. 391–423, the translation on pp. 543–79. There is also a substantial commentary (pp. 139–56). My own supplementary discussion is in Høyrup, 'The "Unknown Heritage"', pp. 632–38.

23 Barthélemy de Romans, *Une arithmétique commerciale du xve siècle*, ed. and trans. by Spiesser, p. 156.

24 Høyrup, 'The "Unknown Heritage"', p. 635, n. 31.

25 Namely Barthélemy de Romans, *Une arithmétique commerciale du xve siècle*, ed. and trans. by Spiesser, pp. 391–402.

Byzantium or in later Greek Antiquity. There is also evidence that Fibonacci himself had learned about that version in Byzantium. His heritage problems speak about *bizantii*, and every time this coin is mentioned together with a place in the *Liber abbaci* (six times in total), this place is Constantinople; the correlation holds the other way too, Byzantium problems never speak about other coin. Either Fibonacci's choice of *bizantii* was simply due to what was in the problem he borrowed, or he wanted to intimate that this was known to him as a Byzantine problem.

However, neither Planudes nor any other Byzantine source knows the sophisticated version, and Fibonacci has no *bizantii* (nor any coin except the almost-abstract *dragma*) in the three corresponding columns. Ibn al-Yāsamīn's downgraded version of the simple problem shows, on the other hand, that this latter version cannot have been part of the general tradition of Arabic recreational mathematics.

That leaves us with only two possible locations for the creation of the sophisticated version: Provence, or the Iberian world, to which Provence had connections. In 2008 I explicitly denied the presence of any version of the problem in the Castilian *Libro de arismética que es dicho alguarismo* from 1393,[26] which in several respects shows more traces of the heritage from al-Andalus than do Catalan and Provençal writings.[27] For that reason I tended to point to Provence, although with strong doubts because of the lack of any evidence that twelfth-century Provence should have been the home of mathematicians with the required level of competence. As it turns out, I was mistaken — both $(1{,}1|\frac{1}{10})$ and $(1{,}1|\frac{1}{11})$ are found on p. 169. On the other hand, no instance of any version is found in known Provençal writings antedating the mid-fifteenth-century *Traicté de la praticque d'algorisme* (which is somehow connected to Barthélemy's *Compendy* and may even be the earlier treatise which he tells to elucidate, cf. n. 21).[28] All in all, an Iberian invention therefore now seems much more plausible than a Provençal origin.

How could the invention have been made? Were the required tools at hand for those who knew to wield them?

Once it is known that a solution can be found, an algebraic solution like the one produced by Fibonacci is, of course, possible. However, it can never provide the idea that such problems should have a solution. It is no doubt feasible to guess, on the basis of the simple version, that other arithmetical series and other fractions will work in more or less the same way, and then try to work it out. But direct proof can also be made by means of a tool that Fibonacci uses elsewhere and must have learned somewhere: the line diagram.

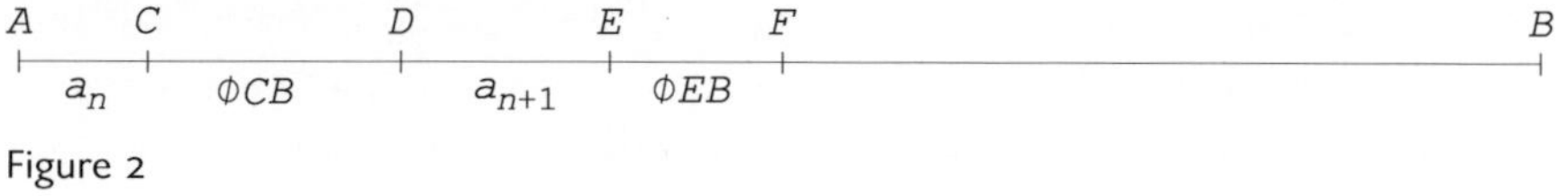

Figure 2

26 *El arte del alguarismo*, ed. by Caunedo de Potro and Córdoba de la Llave, pp. 133–213.

27 Høyrup, 'The "Unknown Heritage"', p. 632.

28 I thank Stéphane Lamassé for giving me access to the transcription of the *Traicté* in his unpublished dissertation.

We may look at Figure 2, supposing that successive shares can be found by taking first absolutely defined amounts a_n, and then a fixed fraction of what remains. *AB* represents S_n, that is, the amount that is available when the *n*-th share is to be taken, *n* being arbitrary (but possible). This share is *AD*, consisting of $AC = a_n$ and $CD = \varphi CB$. The following share is *DF*, consisting of $DE = a_{n+1}$ and $EF = \varphi EB$. Since $AD = DF = \Delta$, $CB = CD+DB$, and $EB = EF+FB$, we find that

$$a_{n+1}-a_n = \varphi(CB-EB) = \varphi(CD-EF)+\varphi(DB-FB) = \varphi(a_{n+1}-a_n)+\varphi\Delta,$$

whence

$$(1-\varphi)\cdot(a_{n+1}-a_n) = \varphi\Delta$$

and further (in order to avoid a formal algebraic division) the proportion

$$\Delta : (a_{n+1}-a_n) :: (1-\varphi) : \varphi\ .$$

Euclid's *Data*, prop. 2 states that: 'If a given magnitude [here Δ] have a given ratio [here $(1-\varphi) : \varphi$] to some other magnitude [here $a_{n+1}-a_n$], the other one is also given in magnitude' (a down-to-earth application of the rule of three is also possible).[29] Therefore, $a_{n+1}-a_n$ is constant, and the successive a_i constitute an arithmetical progression; if they do, on the other hand, all shares will be equal (all steps are linear and therefore reversible).

There can be no doubt that this argument *could* have been made in eleventh-century al-Andalus — and why not then in the twelfth century too, before the final catastrophe?

Liber mahameleth

We shall return to the *Liber abbaci*, but first let us look at the *Liber mahameleth*, an anonymous Latin work that was discovered by Jacques Sesiano in 1974 and first described by him in 1988;[30] in recent years, two critical editions have appeared, prepared respectively by Anne-Marie Vlasschaert and Sesiano.[31] Both editors agree that the title refers to Arabic *al-muʿāmalāt*, which seems indeed beyond dispute. Sesiano proposes that the compilation was made by John of Seville, primarily because a fifteenth-century abbacus book refers to the author as *ispano*, 'Spanish', which Sesiano takes as a mistake for *hispalensis*, 'Sevillian', but also because the *Liber mahameleth* shares a number of passages with John's *Liber algorismi*.[32] He further points to evidence that the work must have been written in an Arabic environment, that is, before John had supposedly gone to Toledo. Sesiano himself raises two

29 Euclid, *Δεδομενα: Euclid's 'Data'*, ed. and trans. by Taisbak, p. 254.

30 *Liber Mahamaleth*, ed. by Sesiano.

31 *Liber Mahamaleth*, ed. by Vlasschaert; *Liber mahameleth* ed. by Sesiano.

32 *Liber mahamaleth*, ed. by Sesiano, pp. xvii–xviii.

objections. Firstly, the Latin in the *Liber mahameleth* is much more polished than in other writings ascribed to John; secondly, the work seems never to have received the final touch, which is at odds with a work produced by somebody who still had many years of activity ahead of him.

On this point, Charles Burnett's work on the identity of John becomes relevant.[33] As it turns out, a group of translations from the Arabic, mostly (perhaps all) from the 1130s and 1140s, are attributed to a Johannes Hispalensis; another group is attributed to one Johannes Hispanus/Hispanis, John of Spain, who was connected to Dominicus Gundisalvi (thus plausibly working in and, in any case, linked to Toledo), who was active around the 1150s, and may but need not be the same person as John of Seville. This was the John who was responsible for the *Liber algorismi*. Burnett points out that the passages which the *Liber mahameleth* shares with the *Liber algorismi* also appear in Gundisalvi's *De divisione philosophiae*, to which it is even closer (Burnett observes further echoes in *De scientiis*, Gundisalvi's translation of al-Fārābī's *On the Classification of the Sciences*). Burnett concludes (p. 70) that the 'impression that one gets, therefore, is that the *Liber mahameleth* is written either by Gundisalvi or by a close associate of his' — and in the latter case Burnett opts for John (of Spain, possibly but not necessarily the same as John of Seville).

Vlasschaert tends to see Gundisalvi as the compiler, though with some reservation.[34] To this an observation can be added which seems to rule out John. The *Liber mahameleth* consistently uses the algebraic terms *census* and *res*, which were to become the standard terminology of Latin algebra (since the systematic introduction to algebra which the *Liber mahameleth* refers to regularly has been lost, there is no occasion to introduce the *radix* as root of the *census*). The second part of the *Liber algorismi*, on the other hand, contains a small 'excerpt of the book called *gebla mucabala*',[35] and here *res*, not *census*, is used as the translation of *māl*. Given the otherwise close relations between the *Liber mahameleth* and the *Liber de algorismi*, this seems to exclude common authorship. More directly it excludes Sesiano's hypothesis that the *Liber mahameleth* was written by John while he was still in Muslim area. Another argument against this hypothesis is that the *Liber de algorismi* makes use of other translations that were circulating in the Toledan area.[36]

However, Sesiano's evidence that the *Liber mahameleth* (or at least the bulk of it) was written in Muslim area remains compelling, while the mathematics of the book clearly betrays the hand of a practising competent mathematician who was at least as familiar with proportion theory as with *al-jabr*, and therefore belonged to the class of mathematicians who integrated mathematics and astronomy.[37] Neither Gundisalvi

---

33 Burnett, 'John of Seville and John of Spain'.

34 *Liber Mahamaleth*, ed. by Vlasschaert, p. 30.

35 'The Toledan *Regule*', ed. by Burnett, Zhao, and Lampe, pp. 163–65.

36 Burnett, 'John of Seville and John of Spain', p. 84.

37 We may speak of these as 'astronomer-mathematicians' — to be distinguished from the class of 'jurisprudent-mathematicians', those who wrote about the kind of mathematics that might be needed by legal scholars — examples of works representing the latter category are ibn Thabāt, *Die Reichtümer der Rechner (Ġunyat al-ḥussāb)*, ed. and trans. by Rebstock; and ʿAlī ibn al-Ḥiḍr, *At-taḏkira bi-uṣūl*

nor his translator friends belonged to that category — their competence in the field, when not mainly metamathematical or philosophical, was rather elementary. Moreover, Gundisalvi's *De divisione philosophiae* mentions a 'book which is called *Mahamalech* in Arabic',[38] and lists its contents in almost exactly the same words as those in the *Liber mahameleth*.[39] In summary, it is therefore almost certain that the *Liber mahameleth* was compiled in Muslim area by an astronomer-mathematician — and almost as certain that the main core of it was not only compiled, but authored. Gundisalvi may still have made amendments — as can be seen if one compares his free translation of al-Fārābī's *On the Classification of the Sciences*, characterized by deletions as well as additions, with the Arabic text and with the translation prepared by Gerard of Cremona.[40]

Of particular interest is the way commercial problems (those belonging with genuine *muʿāmalāt*-mathematics) serve as pretext for advanced experimentation. We may look at those which derived from problems of buying and selling[41]– the ones whose properly practical version would be solved by the rule of three or one of its cognates (those where division precedes multiplication).[42] I omit the detailed calculations for simple variants where, for instance, addition is replaced by subtraction and the operations used for the solution changed correspondingly (for example, from transformation *conjunctim* to *disjunctim*)[43] but no more. Such variation is made systematically throughout the *Liber mahameleth*.

---

al-ḥisāb wa l'farāʾiḍ, ed. and trans. by Rebstock.

38 Dominicus Gundissalinus, *De divisione philosophiae*, ed. by Baur, p. 93. All manuscripts used by Ludwig Baur go back to a single 'probably already secondary and error-ridden' archetype (p. 154). It is quite possible that the original had *Liber mahamaleth*, and that a copyist read a t as c. Conveniently, however, the spelling with *c* allows us to distinguish this Arabic treatise from the Latin compilation.

39 *Liber Mahamaleth*, ed. by Vlasschaert, p. 7.

40 All in al-Fārābī, *Catálogo de las ciencias*, ed. and trans. by González Palencia.

We may thus say that the ascription of the *Liber mahameleth* to John or to Gundisalvi, though made by colleagues who are highly respectful of Arabic science, represents a narrative which tacitly presupposes that only Christians (perhaps Christians living in Muslim area) could get the idea to submit practical knowledge to the inquisitive eye of theory. For a strongly contrasting opinion, cf. Høyrup, 'The Formation of "Islamic Mathematics"'.

41 *Liber Mahameleth*, ed. by Vlasschaert, pp. 193–211.

42 A more complete presentation of this problem group is (supposedly) forthcoming in the proceedings of the '11ième Colloque Maghrébin sur l'histoire des mathématiques arabes, École normale Supérieure, Kouba – Alger, 26, 27, 28 octobre 2013'. Since at that occasion I used Vlasschaert's edition, and since I have also reviewed her volume (Høyrup, review of Vlasschaert ed., *Liber Mahameleth*) and therefore have a heavily annotated copy ready for use, my page references will be to her edition.

43 For ease of reference, also in what follows, I list the full set of operations on a proportion $\frac{a}{b} :: \frac{c}{d}$ as given in Campanus's version of the *Elements*, see Campanus of Novara, [redaction of] *Euclid's 'Elements'*, ed. by H. L. L. Busard, pp. 171–72:

e contrario:	$\frac{b}{a} :: \frac{d}{c}$	conversa:	$\frac{a}{a+b} :: \frac{c}{c+d}$
permutata:	$\frac{a}{c} :: \frac{b}{d}$	eversa:	$\frac{a}{a-b} :: \frac{c}{c-d}$
Conjuncta:	$\frac{a+b}{b} :: \frac{c+d}{d}$	aequa:	$\frac{a}{b} :: \frac{a+c}{b+d}$
disjuncta:	$\frac{a-b}{b} :: \frac{c-d}{d}$		

If p and P stand for prices and q and Q for the appurtenant quantities, we have $\frac{q}{p} :: \frac{Q}{P}$ (this is meant as a proportion, not an equation involving two fractions).[44] The start of each problem is indicated by page and line in the format: 'page$_{\text{line}}$':

193_7 $\frac{3}{13} :: \frac{Q}{P}$, $Q+P = 60$. This is solved by means of proportion theory, namely via transformation into $\frac{3}{3+13} :: \frac{Q}{Q+P}$ and subsequent use of the rule of three.

193_{32} $\frac{3}{13} :: \frac{Q}{P}$, $P-Q = 60$. Similarly.

194_{13} $\frac{3}{8} :: \frac{Q}{P}$, $Q{\cdot}P = 216$. Fractions (or the rules that $\frac{P}{Q} :: \frac{PQ}{QQ} :: \frac{PP}{PQ}$) are not mentioned, but the solution that is offered builds on an awareness that

$$(3{\cdot}216) \div 8 = \tfrac{3}{8}{\cdot}216 = \tfrac{Q}{P}{\cdot}(Q{\cdot}P) = Q^2$$

and

$$(8{\cdot}216) \div 3 = \tfrac{8}{3}{\cdot}216 = \tfrac{P}{Q}{\cdot}(Q{\cdot}P) = P^2$$

194_{27} $\frac{4}{9} :: \frac{Q}{P}$, $\sqrt{Q}+\sqrt{P} = 7\frac{1}{2}$. Uses but does not make explicit that $\frac{\sqrt{4}}{\sqrt{9}} :: \frac{\sqrt{Q}}{\sqrt{P}}$, which is no standard theorem from the theory of proportions[45] but follows easily from an arithmetical understanding. From here as at 193_7.

Alternatively,

$$\sqrt{\tfrac{4}{9}} + 1 = \frac{\sqrt{Q}}{\sqrt{P}} + 1 = \frac{\sqrt{Q}+\sqrt{P}}{\sqrt{P}} = \frac{7\frac{1}{2}}{\sqrt{P}},$$

which also presupposes an underlying arithmetical conceptualization. Yet another alternative makes the claim that

$$\left(\sqrt{\frac{(\sqrt{P}+\sqrt{Q})^2}{(P-Q)/Q} + \left(\frac{\sqrt{P}+\sqrt{Q}}{(P-Q)/Q}\right)^2} - \frac{\sqrt{P}+\sqrt{Q}}{(P-Q)/Q}\right)^2 = Q$$

– which is true but not easy to see or even verify, in particular not if not expressed in modern symbolism. The text does not explain.

195_{17} $\frac{4}{9} :: \frac{Q}{P}$, $\sqrt{P}-\sqrt{Q} = 1\frac{1}{2}$. All three analogous procedures are shown.

196_1 $\frac{4}{9} :: \frac{Q}{P}$, $\sqrt{Q}\cdot\sqrt{P} = 24$. It is tacitly presupposed once again that $\frac{\sqrt{4}}{\sqrt{9}} :: \frac{\sqrt{Q}}{\sqrt{P}}$. The problem is therefore analogous to the one at 194_{13}. However, the first solution that is offered is

$$\frac{\sqrt{P}\cdot\sqrt{Q}}{\sqrt{4}\cdot\sqrt{9}}\cdot 4 = Q, \quad \frac{\sqrt{P}\cdot\sqrt{Q}}{\sqrt{4}\cdot\sqrt{9}}\cdot 9 = P,$$

44 Evidently, this and similar expressions are not proportions in the classical sense (quantities and prices have different dimensions); still, the text not only handles them as if they were but also states explicitly in the beginning of the chapter on buying and selling (p. 186) that the *proportio* of the first quantity to its price is as that of the second quantity to its price. We must think of proportions between the measuring numbers in the given units — then the dimension problem is eliminated.

45 The reverse, when formulated as dealing with the composition of ratios, certainly is.

which suggests an awareness that the initial proportion means that $Q = 4s$, $P = 9s$ with some shared factor s. In the problem at 201_{10} an explicit geometric argument for an analogous insight is given.

Alternatively, a procedure related to that at 194_{13} is suggested.

Finally, it is proposed to multiply 24 by itself, which yields PQ. Then (as explained) the problem is *strictly* analogous to that at 194_{13}.

A chapter follows 'about the same, with [algebraic] things'. When *res* and *census* appear in the text I shall render them here as r and C, respectively.

196_{14} $\frac{3}{10+r} :: \frac{1}{r}$.[46] This is transformed into $\frac{3}{10+r} :: \frac{3}{3r}$, whence $3r = 10+r$, which is solved in the usual *al-jabr* way. Alternatively, the proportion is transformed into $\frac{3-1}{(10+r)-r} :: \frac{1}{r}$, that is, $\frac{2}{10} :: \frac{1}{r}$, whence $\frac{1}{5} :: \frac{1}{r}$, etc. As we can see, cross-multiplication is not used to establish the equation; instead the antecedents are made equal, whence the consequents also become equal. This preference is general.

196_{26} $\frac{4}{20+2r} :: \frac{1½}{2r+3}$. Through multiplication of the right-hand terms by $4 \div 1\frac{1}{2} = 2\frac{1}{3}$, this is transformed into $\frac{4}{20+2r} :: \frac{4}{5\frac{1}{3}r+8}$, whence $5\frac{1}{3}r+3 = 20+r$, etc. Alternatively: $\frac{4}{20+2r} :: \frac{1½}{2r+3} :: \frac{2½}{17}$. But $1½ \div 2½ = \frac{3}{5}$, whence $\frac{1½}{2r+3} :: \frac{1½}{\frac{3}{5}\cdot 17}$, etc. It is pointed out that this ruse only works because we have the same multiple of r left and right.

197_{15} $\frac{8}{20+r} :: \frac{2}{r-1}$. Transforming, we find $\frac{2}{r-1} :: \frac{6}{21} :: \frac{2}{7}$, whence $r-1 = 7$, etc. Alternatively, since $8 \div 2 = 4$, $\frac{8}{20+r} :: \frac{8}{4r-4}$, etc.

197_{33} $\frac{6}{10+r} :: \frac{2}{r}$. Transformed into $\frac{2}{r} :: \frac{4}{10} :: \frac{2}{5}$ whence $r = 5$.

198_{4} $\frac{6}{10+r} :: \frac{2}{r+1}$. By transformation $\frac{2}{r+1} :: \frac{4}{9}$, etc.

198_{14} $\frac{3}{20-r} :: \frac{r/2}{\frac{2}{3}r-2}$ which is transformed into $\frac{3}{20+r} :: \frac{3}{4r-12}$, etc.

198_{24} $\frac{6}{10-r} :: \frac{2}{r}$. By transformation $\frac{2}{r} :: \frac{8}{10}$, whence $\frac{8}{4r} :: \frac{9}{10}$, etc. An alternative that does not depend on the presence of precisely *one* thing left and right transforms the proportion into $\frac{6}{10-r} :: \frac{6}{3r}$, etc.

198_{34} $\frac{4}{8-r} :: \frac{2}{r+1}$. First solved via transformation into $\frac{6}{9} :: \frac{2}{r+1}$, which should give $\frac{2}{3} :: \frac{2}{r+1}$ but by error becomes $\frac{2}{3} :: \frac{2}{r}$, whence $r = 3$. Then, as in the previous example, by the more generally valid alternative, which gives the correct result $r = 2$. The discrepancy is not discussed and thus probably not noticed. Could this be evidence that two different hands were involved?

199_{11} $\frac{4}{20-2r} :: \frac{1½}{2r-3}$. Solved by the 'general' method of the previous two examples. In the end it is pointed out that this can only be understood if one has studied algebra or Euclid's book, 'which however have been sufficiently explained'.

46 'Three measures are given for 10 coins and a thing, but this thing is the price of one measure'. Similarly in the following questions.

In 'another chapter about an unknown in buying and selling' then follows:

199_{27} An unknown number of measures is sold for 93, and adding this number to the price of one measure gives 34 — in our symbols (since no *res* occurs): $x + \frac{93}{x} = 34$. At first the solution is given as $\frac{34}{2} \pm \sqrt{(\frac{34}{2})^2 - 13}$, the sign depending on whether the number of measures exceeds or falls short of the price of one measure. Next, a geometric argument based on the principles of *Elements* II.5 is given. Euclid is not mentioned, however, which the compiler-author is elsewhere fond of doing; since the argument uses a subdivided line only, the direct inspiration might be Abū Kāmil's similar proof for the fifth *al-jabr* case (possession plus number equals things)[47] — elsewhere it is clear that the author knew Abū Kāmil well.

200_2 The first of the two corresponding subtractive variants, namely the one in which the number of measures subtracted from the price of one of them gives 28. First a numerical prescription is given, next a line-based geometric proof. If instead (the second subtractive variant) subtraction of the number of measures from the price of one of them gives 28, one should proceed correspondingly.

201_{10} $\frac{q}{p} :: \frac{Q}{P}$, $pq = 6$, $PQ = 24$, $(p+q)+(P+Q) = 15$. Once again the argument appears to go via the factor of proportionality s, $sp = P$, $sq = Q$ — as this time a geometric argument confirms. At first s is found as $\sqrt{\frac{PQ}{pq}} = 2$. Therefore $p+q = \frac{1}{1+2} \cdot (p+q+P+Q) = \frac{1}{1+2} \cdot 15 = 5$. Since we already know pq, we can proceed according to the fifth case of *al-jabr* or *Elements* II.5, none of which is mentioned; the double solution is, however.

202_{34} $\frac{q}{p} :: \frac{Q}{P}$, $pq = 10$, $PQ = 30$, $(p+q)+(P+Q) = 20$. This seemingly innocuous variation of the preceding question leads to an irrational value $s = \sqrt{3}$, and therefore to complications and a cross-reference to the chapter about roots (where indeed the necessary explanations are found). In the end, this leads to a discussion in terms of the classification of *Elements* X (not mentioned here, which suggests that these classes are supposed to be familiar — elsewhere the book *is* mentioned).

204_{24} $\frac{q}{p} :: \frac{Q}{P}$, $pq = 6$, $PQ = 24$, $(P+Q)-(p+q) = 5$. The first part of this subtractive variant of the problem at 201_{10} is a prescription analogous to the one for the additive variants; for the second part, merely a cross-reference is given.

204_{35} $\frac{q}{p} :: \frac{Q}{P}$, $pq = 6$, $PQ = 24$, $(p+q)\cdot(P+Q) = 10$. Without being identified, the proportionality factor s is found as $\sqrt{\frac{PQ}{pq}}$; next (since $P+Q = s(p+q)$, which also is not explained) $p+q$ is found as $\sqrt{\frac{(p+q)\cdot(P+Q)}{s}}$. For the rest, a cross-reference

47 Abū Kāmil, *Algèbre et analyse Diophantienne*, ed. and trans. by Rashed, pp. 260–63.

is given. For the first step, however, a geometric demonstration is supplied at the end.

205_{21} $\frac{q}{p} :: \frac{Q}{P}$, $pq = 20$, $PQ = 10$, $(p+q)\cdot(P+Q) = \sqrt{5760}$. This is explained to follow the previous question, but evidently gives rise to complicated manipulations of roots, which is why both ways of solving the problem are discussed in detail.

207_{2} $151 = 3q$, $p-q = 34$ (the identification of the two numbers as signifying price and appurtenant quantity is irrelevant). The solution follows from a quadratic completion ($\sqrt{p} = t$, $q = \frac{1}{3}t$):

$$t^2 - \tfrac{1}{3}t = 34$$
$$t^2 - 2\cdot\tfrac{1}{6}t + (\tfrac{1}{6})^2 = 34\tfrac{1}{36}$$
$$T - \tfrac{1}{6} = 5\tfrac{5}{6}$$
$$t = 6$$

At first a purely numerical prescription is given, which is followed by a geometric, line-based proof — that is, the author avoids the automatic procedures of *al-jabr*, apparently preferring what is based on more respectable principles.

207_{24} $\sqrt{p} = 2q$, $p+q = 18$. This additive counterpart of the preceding problem is solved analogously

208_{9} $\frac{6}{4-r} :: \frac{2}{3\sqrt{4+r}}$, which is transformed into $\frac{6}{4+r} :: \frac{6}{9\sqrt{4+r}}$. The resulting equation $(4+r) = 9$ is not stated explicitly, but the numerical prescription corresponds to its transformation into $\sqrt{4+r} = 9$ and further into $4+r = 81$.

208_{15} $\frac{6}{4-r} :: \frac{2}{3\sqrt{4-r}}$. Solved correspondingly.

208_{21} $\frac{3}{x+y} :: \frac{1}{y+\frac{1}{9}y}$, $xy = 21$ (x and y stand for the 'two different things' of the text). The prescription corresponds to the transformation of the proportion into $\frac{3}{x+y} :: \frac{13}{3y+\frac{1}{3}y}$, whence $x+y = 3\frac{1}{3}y$, $x = 2\frac{1}{3}y$, $2\frac{1}{3}y^2 = 21$, $y^2 = 9$, and finally $y = 3$, $x = 7$. After the prescription comes a line-based argument corresponding to these symbolic equations.

Alternatively, the problem can be solved 'according to algebra'. Here, the *thing* (r) takes the place of y, while the *dragma* (d) takes that of x. This time, the equation comes from a different but similar transformation of the proportion, namely into $\frac{1}{\frac{1}{3}d+\frac{1}{3}r} :: \frac{1}{r+\frac{1}{9}r}$. From here follows the equation $r+\frac{1}{9}r = \frac{1}{3}d+\frac{1}{3}r$, and therefore $d = 2\frac{1}{3}r$. Inserting this in $rd = 21$ we get $2\frac{1}{3}C = 21$, $C = 9$, $r = 3$.

209_{30} $\frac{5}{x+y} :: \frac{1}{\frac{1}{3}x+2}$, $xy = 144$. Both methods of the previous problem are applied, now leading to mixed second-degree problems; the line-argument goes through the complete calculation, whereas the algebraic solution satisfies itself with the first step and the explanation that 'the rest is done as we have taught in the algebra'.

The text continues with other classical *muʿāmalāt* topics (profit and interest, partnership, etc.). Also here we first find the proper commercial problems, and afterwards systematic explorations of the sophisticated problems to which the topics can give rise. Obviously, this part of the treatise at least (including the algebra chapter, now lost, to which it refers, and the discussion of roots) was produced as one piece, not put together from disparate books; and obviously, all of it is far beyond the mathematical level of Gundisalvi and his circle, including the John who compiled the *Liber de algorismi*. Nor is it in proper *al-jabr* style, with its ample recourse to proportion techniques and the techniques of *Elements* II — therefore we must suppose the author to have been an astronomer-mathematician presenting *muʿāmalāt*-mathematics *vom höheren Standpunkt aus*, 'from a higher vantage point'.

Back to Fibonacci

We shall now return to the *Liber abbaci*, but to Chapter 15. This chapter is divided into three sections, all relevant to our purpose but most relevant Section 1.[48]

Chapter 15 as a whole is introduced as treating of 'geometrical rules, and questions of algebra and almuchabala', while Section 1 is said to deal with 'proportions of three and four quantities, to which many questions pertaining to geometry are reduced' (p. 387).[49] Actually, the succeeding text consistently talks about *numbers*, not quantities; moreover, the results are not always used in the ensuing 'geometry' section when they would serve. Fibonacci's initial announcement is therefore somewhat out of keeping with what follows.

When three numbers are involved, Fibonacci refers to them as minor/middle/major; as first/middle/major; or as first/second/third. If four, as first/second/third/fourth. For convenience, I shall use $P/Q/R$ or $P/Q/R/S$ respectively here. Mostly, the problems are accompanied by letter-carrying lines drawn in the margin. First come proportions involving three numbers, followed by a few questions involving four numbers. Using conjunction, disjunction, permutation, etc., Fibonacci transforms the given proportion in such a way that the numbers can be found from the product rule (if $\frac{P}{Q} :: \frac{R}{S}$, then $PS = QR$) by means of addition or subtraction or, more often, the principles of *Elements* II.5–6. Fibonacci never refers to Euclid here, as is his habit elsewhere,[50] but only uses line diagrams. Since the omission is systematic, we may already be confident at this point that his use was indirect and the material thus copied; copied, moreover, from a source that did not mention Euclid.

48 I first worked on this in the context of a general investigation of Fibonacci's references to 'proportions', see Høyrup, '"Proportions" in the "Liber abbaci"'.

49 In general, Fibonacci speaks of a 'proportion' (*proportio*) both where we would do so (and where many other medieval authors would write *proportionalitas*) and where we would see a 'ratio'. I shall not try to impose modern distinctions on Fibonacci; even modern usage is indeed inconsistent, failing for instance to distinguish ratios (relations between integers) from fractions (rational numbers). In symbolic language, I shall write the ratio between a and b as $\frac{a}{b}$, corresponding to the way I write proportions, and the corresponding fraction as ${}^{a}/_{b}$.

50 See Article IX in Folkerts, *The Development of Mathematics in Medieval Europe*.

The section can be divided into 50 propositions, most of which are problems[51] — in overview (here # denotes 'number', page numbers are given in superscript; Fibonacci's headings are indicated as '---*heading*---', while divisions '------' correspond to simple paragraph divisions in the edition):

#1–3 deal with three numbers in continued proportion, $P{:}Q{:}R$ (that is, $\frac{P}{Q}::\frac{Q}{R}$) of which one and the sum of the other two are given. The naming of segments presupposes the Latin alphabetic order *a*, *b*, *c*, …

---*Incipit pars prima*---

#1$^{(387)}$ $P+Q=10$, $R=9$. *Conjunctim* $\frac{P+Q}{Q}::\frac{Q+R}{R}$, whence *Elements* II.6 can be applied to $Q\cdot(Q+9)=90$.

#2 $P=4$, $Q+R=15$. Analogous.

#3 $Q=6$, $P+R=13$. The product rule gives $P\cdot R=36$, which is transformed so as to permit the use of *Elements* II.6 (the direct use of II.5 would seem obvious).

#4–38 still consider three numbers, but now differences between the numbers are among the given magnitudes. The alphabetic order changes to *a*, *b*, *g*, *d*, …, pointing to the use of a Greek or an Arabic source. However, in #4–5, still dealing with a continued proportion, *c* is made use of in the calculations:

#4$^{(388)}$ $P{:}Q{:}R$, $Q-P=2$, $R=9$. *Disjunctim* $\frac{R}{Q}::\frac{R-Q}{Q-P}$. Solved by means of *Elements* II.5.

#5 $P{:}Q{:}R$, $R-P=5$, $Q=6$. The product rule gives $P\cdot R=36$, which allows use of *Elements* II.6.

#6 An aside which explains that $\frac{a}{b}::\frac{c}{d}$ entails that the squares of the numbers are also in proportion — a proportion which can then be transformed *conjunctim*, *e converso*, etc. Further, that the same holds for the cubes. This is no consequence of what precedes nor a preparation for what follows immediately (neither a corollary nor a lemma); and when it is eventually used in #50 there is no backwards reference. We may think of note 45, above.

---*Modus alius proportionis inter tres numeros*---

#7$^{(389)}$ $\frac{R-Q}{Q-P}::\frac{R}{P}$, Q unknown. $R-P$ thus has to be split into two parts having the ratio $R{:}P$; this is solved as a partnership problem (the link is not made explicit).

#8 Same proportion, R unknown. *Permutatim* $\frac{R}{R-Q}::\frac{P}{P-Q}$, a first-degree problem.

#9 Same proportion, P unknown, solved similarly.

51 The numbering is mine. Fibonacci has none.

---Modus alius proportionis inter tres numeros---

#10 $\frac{Q-P}{R-Q} :: \frac{R}{P}$, Q unknown. *Conjunctim* $\frac{(Q-P)+(R-Q)}{R-Q} :: \frac{R+P}{P}$, a first-degree problem.

#11(390) Same proportion, *R* unknown. Product rule, and *Elements* II.6.

#12 Same proportion, *P* unknown. Analogously.

---Modus alius proportionis in tribus numeris---

#13 $\frac{R}{P} :: \frac{(R-Q)+(Q-P)}{R-Q}$, Q unknown. Since $(R–Q)+(Q–P) = R–P$, this is a simple first-degree problem.

#14 Same proportion, *R* unknown. $\frac{R-P}{P} :: \frac{Q-P}{R-Q}$. From the product rule follows that the product of *R–P* and *R–Q* is known. So is their difference, which allows application of *Elements* II.6.

#15(391) Same proportion, *P* unknown. Product rule and *Elements* II.5.

---Modus alius proportionis---

#16 $\frac{R}{P} :: \frac{(R-Q)+(Q-P)}{Q-P}$, Q unknown. $\frac{R}{P} :: \frac{R-P}{Q-P}$, a linear problem.

#17 Same proportion, *R* unknown.[52] $\frac{R-P}{P} :: \frac{(R-P)-(Q-P)}{Q-P}$, whence *permutatim* $\frac{R-P}{R-Q} :: \frac{P}{Q-P}$, a linear problem.

#18 Same proportion, *P* unknown. *Eversim* (but Fibonacci writes 'you permutate') $\frac{R}{R-P} :: \frac{R-P}{R-Q}$. The product rule gives *R–P*, whence *P*.

---Incipit differentia tercia in proportione trium numerorum---

#19 No question but the observation that if $\frac{R}{Q} :: \frac{R-Q}{Q-P}$, then *P*, *Q* and *R* are in continued proportion — namely because *Q* must be the same part of *R* as *P* of *Q*. This quasi-arithmetical approach (and the geometric line argument used to support it) is similar to what we encountered in the *Liber mahameleth*, p. 201_{10}.

#20(392) $\frac{R}{Q} :: \frac{Q-P}{R-Q}$, Q unknown. $\frac{Q}{R} :: \frac{R-Q}{Q-P}$, whence $\frac{Q+R}{R} :: \frac{R-P}{Q-P}$. The product rule and an addition allows the use of *Elements* II.6.

#21 Same proportion, *R* unknown. The product rule and *Elements* II.6 give *R*.

#22 Same proportion, *P* unknown. The product rule gives *Q–P*.

---Modus proportionis in tribus numeris---

#23 $\frac{R}{Q} :: \frac{(R-Q)+(Q-P)}{R-Q}$, Q unknown. *Permutatim* and *conjunctim* $\frac{R+(R-P)}{R-P} :: \frac{R}{R-Q}$. From the product rule follows *R–Q*.

52 By error, the text has *minor numerus .a.d.*, but the calculation proceeds from the premise *major numerus .a.b.*, corresponding to *R*.

#24 Same proportion, R unknown. The argument is corrupt (erring in a *convertim* transformation), claiming that the proportion can be transformed into $\frac{Q}{R} :: \frac{R}{Q-P}$.

#25(393) Same proportion, P unknown. $R–P$ follows from the product rule.

---*Modus alius proportionibus in tribus numeris*---

#26 $\frac{R}{Q} :: \frac{R-P}{Q-P}$, Q unknown. *Disjunctim* $\frac{R-Q}{Q} :: \frac{R-Q}{Q-P}$. Since P is a number (i.e., not 0, 'zephirum, hoc est nihil'), R must equal Q. Alternatively, the proportion is transformed *permutatim* into $\frac{R}{R-P} :: \frac{Q}{Q-P}$, and R is posited to be 8, P to be 2, from which is derived that Q must equal R. Finally it is observed that, even with this transformation, the numerical position for P is superfluous.

---*Modus alius proportionis in tribus numeris*---

#27 $\frac{Q}{P} :: \frac{R-Q}{Q-P}$, Q unknown.[53] Instead of transforming *ex aequa* $\frac{Q}{P} :: \frac{Q+(R-Q)}{P+(Q-P)}$, i.e., $\frac{Q}{P} :: \frac{R}{Q}$, Fibonacci prefers to combine transformations *permutatim* $(\frac{R-Q}{Q} :: \frac{Q-P}{P})$ and *conjunctim* $(\frac{Q}{P} :: \frac{R}{Q})$, that is, producing the same outcome. From the product rule, Q is found as $\sqrt{(PR)}$.

#28 Same proportion, R unknown. Fibonacci uses the transformed proportion from #27 to find R as Q^2/P.

#29 Same proportion. It is pointed out that if Q is known (the example being $Q = 12$), then either of the others can be chosen freely, the third number following (via $\frac{Q}{P} :: \frac{R}{Q}$) from division.

---*Modus alius proportionis in tribus numeris*---

#30(394) $\frac{Q}{P} :: \frac{R-P}{Q-P}$, Q unknown. The product rule allows application of *Elements* II.6.

#31 Same proportion, R unknown, $R–P = {}^{Q \cdot [Q-P]}/_{P}$.

#32 Same proportion, P unknown. *Eversim* $\frac{Q}{Q-P} :: \frac{R-P}{R-Q}$. The product rule allows application of *Elements* II.6.

#33 It is then asserted that, if one of the numbers is known in this proportion, the others can be found. What is actually shown (and obviously meant) is that, if one is known, another one can be chosen *ad libitum*, and a third determined so as to fit.

---*Modus alius proportionis in tribus numeris*---

#34 $\frac{Q}{P} :: \frac{Q-P}{R-Q}$, Q unknown. *Conjunctim* $\frac{Q+P}{P} :: \frac{R-P}{R-Q}$, which (via a trick necessitated by the line representation) allows application of *Elements* II.5.

#35 Same proportion, R unknown. $R–Q = {}^{P \cdot [Q-P]}/_{Q}$.

53 The text says *ignotus primus numerus .a.g.*, but *ag* is actually the second, that is, Q.

#36[395] Same proportion, P unknown. The product rule allows application of *Elements* II.5.

#37 $\frac{Q}{P} :: \frac{R-P}{R-Q}$. Since *eversim* $\frac{Q}{Q-P} :: \frac{R-P}{(R-P)-(R-Q)}$, i.e., $\frac{Q}{Q-P} :: \frac{R-P}{Q-P}$, this is only possible if $Q = R-P$ or, as Fibonacci prefers, $P = R-Q$. From this, any one of the numbers can be found if the other two are known.

---*Modus ultimus proportionis in tribus numeris*---

#38 Same proportion, $P+Q+R$ given. For three numbers p, q and r fulfilling the condition, multiply each of them by $^{P+Q+R}/_{p+q+r}$ (a scaling trick that had already been used in Chapter 12, section 2).

#39–50 consider four numbers in proportion, $\frac{P}{Q} :: \frac{R}{S}$. The underlying alphabetic order is still a, b, g, d, …

---*Incipit de proportione quattuor numerorum*---

#39 From $\frac{P}{Q} :: \frac{R}{S}$ follows $\frac{Q}{P} :: \frac{S}{R}$ and $\frac{R}{P} :: \frac{S}{Q}$. From the product rule $PS = QR$, any one of the numbers can be found from the others.

#40 $P+Q$, R and S known. $\frac{P+Q}{Q} :: \frac{R+S}{S}$, whence Q.

#41 $R+S$, P and Q are known. Similarly.

#42 $P+R$, Q and S known. $\frac{Q+S}{S} :: \frac{P+R}{R}$, whence R.

#43 $Q+S$, P and R are known. Similarly.

\------

#44[396] $Q+R$, P and S known. The product rule allows application of *Elements* II.5.

#45 Similarly if $P+S$, Q and R are known. Illustrated by an example involving *rotuli* (a weight unit) and *bizantii* and their sum.

\------

#46 $P-Q$, R and S known. $\frac{P-Q}{Q} :: \frac{R-S}{S}$, whence Q.

#47 $R-S$, P and Q known. Similarly $\frac{P-Q}{Q} :: \frac{R-S}{S}$, whence R.

#48 $P-R$, Q and S known. $\frac{P}{R} :: \frac{Q}{S}$, $\frac{P-R}{R} :: \frac{Q-S}{S}$, whence R.

#49 $P-S$, Q and R known. The product rule allows application of *Elements* II.6.

\------

#50[397] P^2+Q^2, R and S known. Jumps directly (in a numerical example) to the proportion $\frac{P^2+Q^2}{Q^2} :: \frac{R^2+S^2}{S^2}$. Once more, this echoes what we have encountered in the *Liber mahameleth*.

The Latin alphabetic order of #1–3 suggests that this group comes from Fibonacci's own pen, at least in its final redaction. The purely Greek or Arabic order of #7–50 suggests a less edited borrowing. The mixed usage of #4–5 (there is no lettered diagram for #6) probably means that even these were copied, but the calculations reconstructed or made anew by Fibonacci.

The alphabetic order alone only tells us that the source was not Latin; however, the various points of contact of #6–50 with the *Liber mahameleth* suggest the Iberian area and hence al-Andalus.

We shall return to the global project inherent in #7–38 (there is indeed a striking theoretical project); but first we shall look at a few interesting passages from Sections 2 and 3. Section 2 is said to deal with 'questions pertaining to geometry' (p. 397), which again is only partially true.

The questions from Section 2 that concern us here are not geometrical. Instead they deal with commercial travels with composite gain. The first of them (p. 399) is in itself too simple to tell us much — but it provides the background to those that follow. Somebody goes to one place of trade with £100 and earns, and afterwards earns proportionally in another place, and then has a total of £200. A continued proportion (represented by lettered line segments) shows the possession after the first trip to be $\sqrt{(100 \cdot 200)} \approx$ £141, s. 8, d. $5^1/_8$.

The next case (p. 399) is more tricky. The initial capital is still £100 but after the first trip a partner invests £100 in the enterprise, and after the second journey the total amounts to £299. This is expressed in the proportion (represented by lines) $\frac{100}{Q} :: \frac{Q+100}{299}$. From the product rule and *Elements* II.6 (still unidentified) follows the solution $Q =$ £130. An interchange of left and right would reduce this to case #49 from Section 1, but Fibonacci does not make the connection. In any case, the use of proportion theory and the unidentified Euclidean theorem instead of standard algebra[54] indicates a close link to both Section 1 and to the *Liber mahameleth* — a link which is not produced intentionally by Fibonacci, since he does not seem to notice it.

There are two more problems about repeated travels with gain; they also involve manipulation of proportions, but tell us nothing remarkable.

Section 3 is dedicated to 'the solution of certain questions according to the method of algebra and almuchabala, that is, proportion and restoration' (p. 406). This presents us with an enigma. From Gerard's translation of al-Khwārizmī's *Algebra*, Fibonacci knew that one of the techniques of the discipline was 'restoration'.[55] For the other, Gerard's text gave him no certain cues, and he himself only uses *oppositio*/*opponere* (the normal counterparts of *muqabalah*/*qabila*) thrice, and furthermore in the (probably original) sense of confronting the two sides of an equation. Fibonacci therefore had two slots for one operation — and guessed wrongly.[56] But from where did he get the idea that 'proportion' was an essential ingredient? Proportions only play

54 Admittedly, algebra has not been introduced at this point; but nothing asks for these problems to be dealt with in a geometry section.

55 Fibonacci copied from this work and must hence have known it well, see Miura, 'The Algebra in the "Liber abaci" of Leonardo Pisano'.

56 His Arabic, learned presumably during a relatively short period in his boyhood ('some days', he claims in the introduction to the *Liber abbaci*, p. 1) and afterwards perhaps only improved in the course of commercial interactions, must have been less perfect than often assumed.

a minor role in what follows, so a description of his own approach does not seem to be behind the idea. As far as I know, no other obvious explanation is on offer — but if Fibonacci knew something similar to the blend of algebra and proportion theory in the *Liber mahameleth*, he would at least have a reason for his bad guess.

Among the appearances of proportions within the section, at least one seems to belong to the types we know from Section 1 and from the *Liber mahameleth* (others are possible but not sufficiently characteristic). On p. 423, Fibonacci asks for a number which, when $^1/_3$ of it and 6 more are removed and the remainder multiplied by itself, yields twice the original number — in symbols,

$$(x-{}^1/_3x-6)^2 = 2x\,.$$

We are told that this could be found by algebra, but that is not done. Instead, Fibonacci makes use of a line diagram and transforms the data into a proportion which in symbols becomes

$$\frac{\frac{2}{3}x}{x-\frac{1}{3}x-6} :: \frac{x-\frac{1}{3}x-6}{3}\,.$$

Disjunctim, this allows him to apply *Elements* II.6 (unidentified once again). The underlying alphabetic order is *a, b, g, d*.

So, Sections 2 and 3 contain scattered traces of that same influence which dominates Section 1, and which has a remarkable similarity to what we know from the *Liber mahameleth*. But, as promised, we shall have another look at Section 1, whose #7–38 present us with an unparalleled theoretical investigation.

Pappos's *Collection* III, 'Concerning plane and solid geometrical problems',[57] largely deals with *means*. Propositions xii–xxiii (pp. 70–105), in particular, deal with a set of 10 means between two magnitudes — the original three (arithmetical, geometrical, harmonic), which are 'very useful for dealing with ancient writings'; with three more that were added by early authors, and finally with four joined to them by 'the more recent' (pp. 84–85). The last part of Nicomachos's *Introduction to Arithmetic* deals with almost the same, though now between numbers.[58] The two are discussed in parallel by Thomas L. Heath, who observes that one of Nicomachos's means is omitted by Pappos, and vice versa.[59]

As it turns out, the sequence #7–38 is very closely related to this ancient classification and theory of means: for two numbers *P* and *R*, and the mean *Q*, it is shown how knowledge of any two of them allows us to find the third. The scheme on top of the following page illustrates it. I have followed Nicomachos's order, as being the most relevant. As we see, indeed, Fibonacci agrees with Nicomachos and Boethius but not with Pappos in the cases 4–6, having $\frac{R}{P} :: \frac{Q-P}{R-Q}$ instead of $\frac{R-Q}{Q-P} :: \frac{P}{R}$, etc. We may

57 Pappos of Alexandria, *Collectionis quae supersunt*, ed. by Hultsch, vol. 1, pp. 30–177.

58 Nicomachos of Gerasa, *Introductionis arithmeticae libri II*, ed. by Hoche, pp. 122–44; Nicomachos of Gerasa, *Introduction to Arithmetic*, ed. by D'Ooge, pp. 266–84.

59 Heath, *A History of Greek Mathematics*, pp. 86–89.

	Pappos	Nicomachos	*Liber abbaci*
$\frac{R-Q}{Q-P} :: \frac{R}{R}$ (arithmet.)	P1	*N1*	
$\frac{R-Q}{Q-P} :: \frac{R}{Q}$ or $\frac{R-Q}{Q-P} :: \frac{Q}{P}$	P2	*N2*	*#27–29*
$\frac{R-Q}{Q-P} :: \frac{R}{P}$	P3	*N3*	*#7–9*
$\frac{R-Q}{Q-P} :: \frac{P}{R}$	P4	*N4 (but inverted)*	*#10–12 (inverted)*
$\frac{R-Q}{Q-P} :: \frac{P}{Q}$	P5	*N5 (but inverted)*	*#34–36 (inverted)*
$\frac{R-Q}{Q-P} :: \frac{Q}{R}$	P6	*N6 (but inverted)*	*#20–22 (inverted)*
$\frac{R-P}{Q-P} :: \frac{R}{P}$	absent	*N7*	*#16–18*
$\frac{R-P}{R-Q} :: \frac{R}{P}$	P9	*N8*	*#13–15*
$\frac{R-P}{Q-P} :: \frac{Q}{P}$	P10	*N9*	*#30–32*
$\frac{R-P}{R-Q} :: \frac{Q}{P}$	P7	*N10*	*#37–38*
$\frac{R-P}{R-Q} :: \frac{R}{Q}$	P8	*absent*	*#23–25*

thus assume that the ultimate inspiration is Nicomachos, not Pappos. This is not astonishing, Nicomachos was well known in both the Arabic and the Latin worlds (in the latter in Boethius's translation).[60] My initial quotation from Djebbar also pointed to a tradition starting from Thābit ibn Qurra's translation of Nicomachos.

There are obvious differences, however. Firstly, the *Liber abbaci* contains no counterpart of the arithmetical mean, whose expression as a proportion is indeed next to ridiculous — the only sensible definition is $R–Q = Q–P$; whoever made Nicomachos's material the object of systematic theoretical exploration has seen that. On the other hand, the case omitted by Nicomachos but discussed by Pappos is included. This is not evidence of contamination, a competent mathematician going through all the cases would observe that it had its place. Similarly he would observe that Fibonacci's #26 ought to be there, at least once we have given up the idea that Q is a genuine mean and therefore $P<Q<R$; and as we see, Fibonacci does not speak of means (his reference to the 'middle' number is only occasional, and serves as identification only), for which reason we may assume that his source did not either. Finally, the treatment of means is extended by a similar investigation of

60 Boethius's presentation agrees exactly with Nicomachos in all relevant respects, see Boethius, *De institutione arithmetica libri duo*, ed. by Friedlein, pp. 140–69.

four numbers in proportion, which — because of the alphabetic ordering — is also likely to represent a borrowing, plausibly from the same source.

All in all: the vicinity of the methods of Fibonacci's #7–50 to those of the *Liber mahameleth*, together with the uniqueness of the two texts, makes it reasonably certain that Fibonacci's source came, if not from the same hand then at least from the same environment as the theoretical exploration of the possibilities offered by *muʿāmalāt* mathematics.

On the other hand, the absence of anything in known Ibero-Latin texts similar to Fibonacci's Chapter 15 Section 1 makes it implausible that either of the two emerged from the ambience of translators into Latin and the users of their translations — even more implausible, we may say, than the idea is in itself that the *Liber mahamalech* was produced by Gundisalvi or an associate of his.

Jacopo's Quasi-Algebra

A vernacular source of interest is Jacopo da Firenze's *Tractatus algorismi* — more precisely a sequence of problems dealing with the wages of the manager of a *fondaco* over three or four years, supposed to increase geometrically.[61] Designating the four salaries a, b, d and e, we may express the problems thus:

(1) $a+d = 20$, and $b = 8$.

(2) $a = 15$, $e = 60$.

(3) $a+e = 90$, $b+d = 60$.

(4) $a+d = 20$, $b+e = 30$

All solutions are given as numerical prescriptions only, there is neither *cosa-censo* algebra nor line diagrams (the latter are never used by Jacopo). But it is easy to change the prescriptions unambiguously into formulae.

Problem (1) is solved in this way:[62]

$$a = \tfrac{a+d}{2} - \sqrt{\left(\tfrac{a+d}{2}\right)^2 - ad} \quad \text{and} \quad d = \tfrac{a+d}{2} + \sqrt{\left(\tfrac{a+d}{2}\right)^2 - ad}.$$

Obviously, the product rule gives $ad = b^2$, after which the rule corresponding to *Elements* II.5 gives a and d.

61 Once more only in the Vatican manuscript Vat. lat. 4826, Jacopo da Firenze. *Tractatus Algorismi*, ed. by Høyrup, pp. 326–31, where it is separated from the algebra section proper by only an alligation problem dealing with grain; cf. above, note 9.

62 I discover a misprint in my analysis in Jacopo da Firenze, *Tractatus Algorismi*, ed. by Høyrup, p. 116 — in the formula to the right, b should be d (as here).

Problem (2) is solved as

$$b=\sqrt[3]{\tfrac{d}{a}\cdot a^3}\ ,\ \dots\ d=\sqrt[3]{\left(\tfrac{d}{a}\right)^2\cdot a^3}\ ,$$

which just requires familiarity with the theory of continued proportions (duly generalized so as to accept irrational radicals). If we think in terms of a factor of proportionality *s*, as we know it from the *Liber mahameleth*, it can be solved as the algebraic problem 'cubes equal number', for which Jacopo has presented the rule earlier on[63] — but no such link is made, and it is doubtful whether Jacopo really understood algebra beyond the second degree (he offers no examples, only rules).

Problem (3) is more surprising. It is solved by means of the formula

$$a{\cdot}e=b{\cdot}d=\frac{(b+d)^3}{3(b+d)+(a+e)}\,,$$

which is easily justified if we think in terms of a factor of proportionality *s*, since then

$$\frac{(b+d)^3}{3(b+d)+(a+e)}=\frac{a^3p^3(1+p)^3}{a(3p+3p^2+1+p^3)}=a^2p^3=a\cdot ap^3=ap\cdot ap^2\,.$$

How it was actually found is another matter. In any case, once $a{\cdot}e=b{\cdot}d$ are determined, $a{+}e$ and $b{+}d$ being already given, all four can be found (and are indeed found) by means of the rule corresponding to *Elements* II.5.

Without identifying it, problem (4) finds the factor of proportionality p as $(b{+}e)/(a{+}d)$, whence

$$a=\tfrac{a+d}{1+p^2},\ d=(a{+}d)-a\ ,\ b=\tfrac{b+e}{1+p^2},\ e=(b{+}e)-b\ .$$

Since the factor of proportionality is actually found and used here, we may be confident that it also underlies the solutions to problems (2) and (3). Together with the use of a rule derived from *Elements* II.5, already this suggests kinship with *Liber mahameleth* and *Liber abbaci*, chapter 15. The idea of using a (pseudo)practical problem as a starting point or pretext for deeper study, on the other hand, recalls what was done to the unknown heritage and the treatment of commercial problems in the *Liber mahameleth*. In the absence of anything similar in earlier extant sources[64] we may therefore assume with fair certainty that even this group of problems was drawn from the same al-Andalus reservoir as the three Latin sources discussed above.

Nothing is certain in this world. There are even proponents of the theory that Shakespeare's works were not written by Shakespeare himself but by somebody else also called Shakespeare, as we know. But with that proviso I think we may conclude

63 Jacopo da Firenze, *Tractatus Algorismi*, ed. by Høyrup, p. 320.

64 Scattered later sources until Pacioli, Tartaglia and Cardano do contain some similar problems, with variations that suggest Jacopo not to have been the only channel through whom the problem type reached the Italian abbacus environment — see Jacopo da Firenze. *Tractatus Algorismi*, ed. by Høyrup, pp. 117–20 and Kichenassamy, 'Continued Proportions and Tartaglia's Solution of Cubic Equations'.

that twelfth-century al-Andalus was the home to astronomer-mathematicians who systematically developed *theory* or something close to that from the questions of simpler, non-theoretical mathematics — be it the puzzling unknown heritage, be it commercial arithmetic, be it Nicomachos's list of means, be it finally the riddle about geometrically-increasing wages.

Bibliography

Primary Sources

Abū Kāmil, *Algèbre et analyse Diophantienne*, ed. and trans. by Roshdi Rashed (Berlin: de Gruyter, 2012)

al-Fārābī, *Catálogo de las ciencias*, Segunda edición, ed. and trans. by Ángel González Palencia (Madrid: CSIC, Instituto Miguel Asín, 1953)

ʿAlī ibn al-Ḥiḍr, *At-taḏkira bi-uṣūl al-ḥisāb wa l'farāʾiḍ (Buch über die Grundlagen der Arithmetik und der Erbteilung)*, ed. and trans. by Ulrich Rebstock (Frankfurt a.M.: Institute for the History of Arabic-Islamic Science at the Johann Wolfgang Goethe University, 2001)

Barthélemy de Romans, *Une arithmétique commerciale du* XV*e siècle. Le 'Compendy de la praticque des nombres'*, ed. and trans. by Maryvonne Spiesser (Turnhout: Brepols, 2003)

Boethius, *De institutione arithmetica libri duo. De institutione musica libri quinque. Accedit Geometria quae fertur Boetii*, ed. by Gottfried Friedlein (Leipzig: Teubner, 1867)

Campanus of Novara, [redaction of] *Euclid's 'Elements'*, 2 vols, ed. by H. L. L. Busard (Stuttgart: Franz Steiner, 2005)

Cardano, Girolamo, *Practica arithmetice, et mensurandi singularis* (Milano: Bernardini Calusco, 1539)

El arte del alguarismo. Un libro castellano de aritmética comercial y de ensayo de moneda del siglo XIV (MS. 46 de la Real Colegiato de San Isidoro de León), ed. by Betsabé Caunedo del Potro and Ricardo Córdoba de la Llave (Salamanca: Junta de Castilla y León, Consejería de Educación y Cultura, 2000)

Euclid, *Δεδομενα: Euclid's 'Data', or, The Importance of Being Given*, ed. and trans. by Christian Marinus Taisbak (Copenhagen: Museum Tusculanum, 2003)

Euler, Leonhard, *Élémens d'algebre* I. *De l'analyse déterminée* II. *De l'analyse indéterminée* (Lyon: Jean-Marie Bruyset, 1774)

Fibonacci, *Il Liber abbaci, Scritti di Leonardo Pisano matematico del secolo decimoterzo*, I, ed. by Baldassare Boncompagni (Rome: Tipografia delle Scienze Matematiche e Fisiche, 1857)

Gundissalinus, *De divisione philosophiae*, ed. by Ludwig Baur, Beiträge zur Geschichte der Philosophie des Mittelalters. Texte und Untersuchungen, 4. 2–3 (Münster: Aschendorffsche Buchhandlung, 1903)

Heron of Alexandria, *Definitiones cum variis collectionibus. Heronis quae feruntur Geometrica*, ed. and trans. by J. L. Heiberg, Heronis Alexandrini Opera quae supersunt omnia, 4 (Leipzig: Teubner, 1992)

Jacopo da Firenze, *Tractatus Algorismi and Early Italian Abbacus Culture*, ed. by Jens Høyrup (Basel: Birkhäuser, 2007)

Liber mahameleth, ed. by Anne-Marie Vlasschaert (Wiesbaden: Franz Steiner, 2010)

Liber mahameleth: A Twelfth-Century Mathematical Treatise, ed. by Jacques Sesiano (Heidelberg: Springer, 2014)

Maximos Planudes, *Le Grand calcul selon les Indiens*, ed. and trans. by André Allard (Louvain-la-Neuve: Faculté de Philosophie et Lettres, 1981)

Nicomachos of Gerasa, *Introduction to Arithmetic*, ed. by Martin Luther D'Ooge (London: MacMillan, 1926)

Nicomachos of Gerasa, *Introductionis arithmeticae libri II*, ed. by Richard Hoche (Leipzig: Teubner, 1866)

Paolo Dell'Abaco, *Trattato d'aritmetica*, ed. by Gino Arrighi (Pisa: Domus Galilæana, 1964)

Pappos of Alexandria, *Collectionis quae supersunt*, I, II, III.1–2, ed. by Friedrich Hultsch (Berlin: Weidmann, 1876–1878)

Ibn Thabāt, *Die Reichtümer der Rechner (Ġunyat al-ḥussāb). Die Araber — Vorläufer der Rechenkunst*, ed. and trans. by Ulrich Rebstock (Walldorf-Hessen: Verlag für Orientkunde Dr H. Vorndran, 1993)

'The Toledan *Regule* ("Liber Alchorismi", part II): A Twelfth-Century Arithmetical Miscellany', ed. by C. Burnett, J.-W. Zhao, and K. Lampe, *SCIAMUS*, 8 (2007), 141–231

Secondary Studies

Burnett, Charles, 'John of Seville and John of Spain, a *mise au point*', *Bulletin de philosophie médiévale*, 44 (2002), 59–78

Djebbar, Ahmed, 'Deux mathématiciens peu connus de l'Espagne du XIe siècle: al-Maʾtaman and Ibn Sayyid', in *Vestigia Mathematica: Studies in Medieval and Early Modern Mathematics in Honour of H. L. L. Busard*, ed. by M. Folkerts and J. P. Hogendijk (Amsterdam: Rodopi, 1993), pp. 79–91

Folkerts, Menso, *The Development of Mathematics in Medieval Europe: The Arabs, Euclid, Regiomontanus* (Aldershot: Ashgate, 2006)

Heath, Thomas L., *A History of Greek Mathematics*, 2 vols (Oxford: Clarendon Press, 1921)

Høyrup, Jens, 'The Formation of "Islamic Mathematics": Sources and Conditions', *Science in Context*, 1 (1987), 281–329

———, 'The "Unknown Heritage": Trace of a Forgotten Locus of Mathematical Sophistication', *Archive for History of Exact Sciences*, 62 (2008), 613–54

———, '"Proportions" in the "Liber abbaci"', in *Proportions: Science – Musique – Peinture & Architecture: Actes du LIe Colloque International d'Études Humanistes, 30 juin – 4 juillet 2008*, ed. by Sabine Rommevaux, Philippe Vendrix, and Vasco Zara (Turnhout: Brepols, 2011), pp. 83–100

———, Review of Anne-Marie Vlasschaert ed., *Liber Mahameleth* (2010), *Speculum*, 90 (2015), 1177–78

———, 'Which Kind of Mathematics Was Known and Referred to by Those Who Wanted to Integrate Mathematics in "Wisdom" — Neopythagoreans and Others?', *AIMS Mathematics*, 1 (2016), 24–42

Kichenassamy, Satyanad, 'Continued Proportions and Tartaglia's Solution of Cubic Equations', *Historia Mathematica*, 42 (2015), 407–35

Lorch, Richard P., 'Jābir ibn Aflaḥ', *in Dictionary of Scientific Biography*, vol. 7 (New York: Scribner, 1973), pp. 37–39

Miura, Nobuo, 'The Algebra in the "Liber abaci" of Leonardo Pisano', *Historia Scientiarum*, 21 (1981), 57–65

Sabra, Abdelhamid I., 'The Appropriation and Subsequent Naturalization of Greek Science in Medieval Islam: A Preliminary Statement', *History of Science*, 25 (1987), 223–43

Sesiano, Jacques, 'Le *Liber Mahamaleth*, un traité mathématique latin composé au XIIe siècle en Espagne', in *Histoire des Mathématiques Arabes. Premier colloque international sur l'histoire des mathématiques arabes, Alger, 1.2.3 décembre 1986. Actes* (Alger: La Maison des Livres, 1988), pp. 69–98

Note added in proof:
The 'enigma' discussed on pp. 54–55 turns out to be a red herring not asking for explanation. Enrico Giusti's new critical edition of the *Liber abbaci* (Firenze: Olschki, 2000) reveals that the explanation of *algebra et almuchabala* as 'ad proportionem et restaurationem' is a mistake in the manuscript used by Boncompagni. All other manuscripts have 'appositionis et restaurationis'.

ALEXANDER FIDORA

Gundissalinus, Arabic Philosophy, and the Division of the Sciences in the Thirteenth Century

*The Prologues in Philosophical Commentary Literature**

During the twelfth century, Arabic philosophy and science strongly influenced the Latin West through the translations from Arabic into Latin that were prepared in the city of Toledo and elsewhere. One of the most outstanding figures in this process of transmission was Dominicus Gundissalinus (1110–1190) who not only produced more than twenty translations of fundamental philosophical texts by authors such as al-Kindī and Avicenna, but also composed works of his own that rely heavily on the materials he translated.[1]

Among Gundissalinus's own works, his philosophical encyclopaedia *De divisione philosophiae* deserves a special place. It can be considered a hallmark in the history of philosophy, because this synthesis of Latin and Arabic learning, which draws strongly on al-Fārābī's *Kitāb iḥṣāʾ al-ʿulūm*, introduces a number of hitherto unknown sciences into Latin philosophy, such as optics (*de aspectibus*) and statics (*de ponderibus*), while substantially redefining others, namely metaphysics and politics.[2]

While today the importance of Dominicus Gundissalinus and his division of philosophy is generally acknowledged, the significance of his contribution for later epistemological developments still needs to be assessed

* I thank John N. Crossley (Melbourne) for reading a draft of this chapter.

1 See Hasse, 'Influence of Arabic and Islamic Philosophy on the Latin West'.

2 See Fidora, 'Dominicus Gundissalinus and the Introduction of Metaphysics into the Latin West'.

Alexander Fidora • (alexander.fidora@icrea.cat) is an ICREA research professor in the Department of Ancient and Medieval Studies of the Universitat Autònoma de Barcelona. His research focuses on medieval philosophy as well as the intercultural and interreligious dimensions of medieval thought. He has directed the ERC Starting grant 'Latin into Hebrew' (2008–2012) and the ERC Consolidator project 'The Latin Talmud' (2014–2019).

Premodern Translation: Comparative Approaches to Cross-Cultural Transformations, ed. by Sonja Brentjes and Alexander Fidora, Contact and Transmission, 2 (Turnhout: Brepols, 2021), pp. 63–88

10.1484/M.CAT-EB.5.122134

with more precision. For a long time, histories of philosophy provided only a few references to Michael Scot (d. before 1236) and his tract *De divisione philosophiae*, along with Robert Kilwardby (d. 1279) and his famous *De ortu scientiarum*, when dealing with the influence of Gundissalinus's division of the sciences. In recent years, however, attention has been drawn to a very prolific literary genre that flourished at the Faculty of Arts of the University of Paris in the middle of the thirteenth century, the so-called introductions to philosophy, that is, treatises on the division of the sciences in which the masters articulated how they understood their profession as philosophers.[3] This important corpus proves that Gundissalinus's influence extended much further, his epistemological account playing a pivotal role at the Faculty of Arts during the 1240s and 1250s, when philosophy established itself as an autonomous discipline at the University of Paris.[4] As I have shown elsewhere, the Arabic classification of the sciences that Gundissalinus transmitted offered a series of systematic criteria to the authors of the introductions that were instrumental in defining the subject matter of philosophy and determining its overall structure as well as its subdivisions.[5]

In this paper I shall show that, also during the following decades, when the genre of introductory treatises to philosophy declined, Gundissalinus's division of philosophy persisted in the philosophical discourse. The introductions had developed in close connection with another genre, namely the prologues in philosophical commentary literature, and these prologues, for their part, continued using epistemological distinctions from Gundissalinus in their attempt to describe the specific place of the works to be commented on in the *ordo scientiarum*. Consequently, Gundissalinus's division of philosophy served as a clear frame for the reading of Aristotle and other philosophers, even long after the genre of introductory treatises to philosophy had fallen into disuse.

While the connection between the introductions to philosophy on the one hand and the epistemological reflections in philosophical commentary literature on the other was already noted by Martin Grabmann,[6] Gundissalinus's specific role in the prologues to the commentaries has not been hitherto addressed. In what follows, I shall therefore trace Gundissalinus's influence on two prologues from the first half of the thirteenth century until the later expositions by Peter of Ireland and Thomas Aquinas in the second half of the thirteenth and John of Jandun in the fourteenth century.

---

3 Many of these texts have been edited recently thanks to the efforts of Claude Lafleur and others. See Lafleur, *Quatre Introductions à la philosophie au XIIIe siècle*. Cf. also Imbach, 'Einführungen in die Philosophie aus dem XIII. Jahrhundert'.

4 Cf. Gorochov, *Naissance de l'Université*, pp. 555–73.

5 See Fidora, 'The Arabic Influence on the Classification of Philosophy in the Latin West'.

6 See Grabmann, 'Die logischen Schriften des Nikolaus von Paris', pp. 241–42.

Two Introductions to the *Isagoge* From the First Half of the Thirteenth Century

The first text that deserves our attention is the anonymous introduction to the study of Porphyry's *Isagoge* that Gilbert Dahan has edited under the title *Ut ait Tullius*.[7] The sources and doctrines of this text suggest a date slightly before the year 1250, particularly if compared to other mid-thirteenth-century epistemological accounts, which articulate a more Aristotelian concept of philosophy. From an epistemological point of view, two sections in this text deserve attention, namely the paragraphs on the definition of philosophy and those devoted to its division. Most revealing with regard to Gundissalinus's influence is the former, for, as Gilbert Dahan has shown, *Ut ait Tullius* draws strongly on the following list of definitions from the prologue to Gundissalinus's *De divisione philosophiae*:

1. Philosophia est assimilatio hominis operibus creatoris secundum virtutem humanitatis (Philosophy is the assimilation of man to the works of the creator according to the power of man).
2. Philosophia est taedium et cura et studium et sollicitudo mortis (Philosophy is disgust [of life] and concern, aspiration and anxiety for death).
3. Philosophia est rerum humanarum divinarumque cognitio cum studio bene vivendi (Philosophy is the cognition of human and divine things joined with a zeal for living well).
4. Philosophia est ars artium et disciplina disciplinarum (Philosophy is the art of arts and the discipline of disciplines).
5. Philosophia est integra cognitio hominis de se ipso (Philosophy is man's total cognition of himself).
6. Philosophia est amor sapientiae (Philosophy is the love of wisdom).[8]

At least three out of seven definitions in *Ut ait Tullius* can be traced back to Gundissalinus, namely:

> Philosophia est amor sapientiae, et dicitur a *phylos* quod est *amor*, et *sophos* quod est *sapientia*. Unde antiquitus philosophi solebant dici per arrogantiam; nomine tamen philosophiae nuncupantur, quasi amatores philosophiae.
>
> (Philosophy is the love of wisdom, and its name comes from *phylos* which means *amor*, and *sophos* which means *wisdom*. Thus, in ancient

7 Dahan, 'Une introduction à l'étude de la philosophie: *Ut ait Tullius*'.

8 Dominicus Gundissalinus, *De divisione philosophiae*, ed. by Fidora and Werner, pp. 56–60. As Gundissalinus himself compiles these definitions from various traditional sources, notably from Isaac Israeli's *Liber de definitionibus*, it is not always easy to establish a direct dependence between our text and his *De divisione philosophiae*. Yet, Claude Lafleur's edition of the introductions to philosophy has shown that many of the definitions of philosophy reached the Parisian masters through Gundissalinus. See, for a doxographical analysis of Gundissalinus's list of definitions, Chroust, 'The Definitions of Philosophy in the *De divisione philosophiae*'. For the English translation of nos 1 and 3–6, see Grant, *A Source Book in Medieval Science*, p. 60.

times, one spoke of philosophers not without some pretention; yet they are called by the name of philosophy insofar as they are lovers of philosophy.)[9]

As Gilbert Dahan has noticed, this passage goes back to the sixth definition from the prologue to Gundissalinus's *De divisione philosophiae*, which in its entirety reads as follows:

> Philosophia est amor sapientiae. *Philos* enim graece *amor* dicitur latine; et *sophia* dicitur *sapientia*; inde philosophia est amor sapientiae et philosophus dicitur amator sapientiae.
>
> > (Philosophy is the love of wisdom. For *philos* in Greek is translated *love* in Latin and *sophia* is translated *wisdom*. Whence philosophy is the love of wisdom, and a philosopher is called lover of wisdom.)[10]

Next, the author of *Ut ait Tullius* puts forward a definition that he attributes to Aristotle and al-Ġazālī. Yet, as Dahan argues, this definition seems to echo again Gundissalinus and his encyclopaedia.[11] It reads as follows:

> Ab Aristotele sic definitur philosophia: 'Philosophia est cognitio sui ipsius ab homine'. Unde Algazel: 'Cogoscente te ipsum, et cognosces omnia, quia in te quodammodo sunt omnia'. Omne autem, secundum quod dicit Aristoteles, aut est substantia aut accidens.
>
> > (Aristotle defines philosophy as follows: 'Philosophy is man's cognition of himself'. Therefore al-Ġazālī says: 'Knowing yourself, you will know everything, for in a certain sense there is everything in you'. But according to Aristotle, everything is either a substance or an accident.)[12]

The passage draws on the fifth of Gundissalinus's above definitions, which reads in full:

> Philosophia est integra cognitio hominis de se ipso. Cum autem homo cognoscit se ipsum integre, profecto cognoscit quicquid est. In homine enim sunt substantia et accidens. Sed substantia est duplex: scilicet spiritualis, sicut anima et intelligentia, et corporea, sicut corpus longum, latum et spissum. Similiter et accidens duplex est, quia spirituale et corporale.

---

9 Dahan, 'Une introduction à l'étude de la philosophie: *Ut ait Tullius*', p. 43.

10 Dominicus Gundissalinus, *De divisione philosophiae*, ed. by Fidora and Werner, pp. 58/60; English translation in Grant, *A Source Book in Medieval Science*, p. 60. Gundissalinus himself is indebted to Isaac Israeli for this definition; yet the wording in *Ut ait Tullius* is closer to *De divisione philosophiae* than to the *Liber de definitionibus*. See below.

11 As Dahan, 'Une introduction à l'étude de la philosophie: *Ut ait Tullius*', states, many attributions in *Ut ait Tullius* are very imaginative (p. 8) and require thorough examination (pp. 18–19).

12 Dahan, 'Une introduction à l'étude de la philosophie: *Ut ait Tullius*', p. 44.

> > (Philosophy is man's total cognition of himself. This is because when a man recognizes himself completely, indeed he recognizes everything that exists. For in man there are substance and accident, but substance is twofold: namely spiritual, like soul and intelligence, and corporeal, like a long and wide and thick body. Similarly, accident is twofold: spiritual and corporeal.)[13]

The third definition that deserves our attention is particularly telling, being the well-known description of philosophy from Isidore of Seville's *Etymologiae*. II. 24. 1., which is, however, attributed once more to al-Ġazālī:

> Algazel autem sic definit ipsam: Philosophia est divinarum humanarumque rerum cognitio.
>
> > (al-Ġazālī, however, defines it as follows: Philosophy is the cognition of divine and human things.)[14]

That the anonymous author of *Ut ait Tullius* associates this definition with an Arabic author, instead of Isidore, proves that he is not working with the original sources for his list, but rather has come across the definitions in Gundissalinus's treatise, which, due to its strong reliance on Arabic materials, was frequently attributed to Arabic authors.[15]

The author of *Ut ait Tullius* is not the only contemporary to quote Isidore's definition from Gundissalinus's treatise. In his important introduction to philosophy, the *Divisio scientiarum*, written in about 1250, Arnulf of Provence includes among his twelve definitions of philosophy the following reference:

> Secunda [definitio] habetur in libro *De divisione philosophiae*, quae est talis: Philosophia est divinarum humanarumque rerum certa cognitio.
>
> > (The second definition occurs in the book *On the Division of Philosophy*, namely: Philosophy is the certain cognition of divine and human things.)[16]

---

13 Dominicus Gundissalinus, *De divisione philosophiae*, ed. by Fidora and Werner, p. 58; English translation from Grant, *A Source Book in Medieval Science*, p. 60. Also for this definition, Gundissalinus has recourse to Isaac Israeli's *Liber de definitionibus*, which was translated twice into Latin, by Gerard of Cremona and Dominicus Gundissalinus himself. Gerard's translation differs considerably from the wording in *Ut ait Tullius*, while Gundissalinus's translation is very close to both *Ut ait Tullius* and his treatise *De divisione philosophiae*. See Isaac Israeli, *Liber de definitionibus*, ed. by Muckle, p. 306 (= trans. Gerard) and 331 (= trans. Gundissalinus).

14 Dahan, 'Une introduction à l'étude de la philosophie: *Ut ait Tullius*', p. 45.

15 See, for instance, John Dee's manuscript of the work, which ends as follows: 'Explicit liber Gundisalvi *De divisione philosophiae*; alii putant quod sit Alpharabii' (Oxford, Bodleian Library, MS Digby 76, fol. 107v). The page with the explicit is reproduced in Dominicus Gundissalinus, *De divisione philosophiae*, ed. by Fidora and Werner, p. 271.

16 Lafleur, *Quatre Introductions à la philosophie au* XIII*e* *siècle*, p. 308.

To this he adds:

> Haec plenius et completius ponitur ab Isidoro, in tertio *Etymologiarum*, ubi adiungitur: Philosophia est divinarum humanarumque etc. cum studio bene vivendi coniuncta.
>
> > (The same is given in an extended and more complete form by Isidore, in the third book of his *Etymologies*, where the following is added: Philosophy is [the cognition] of divine and human etc. joined with a zeal for living well.)[17]

While Arnulf's references are much more accurate than the one from *Ut ait Tullius*, they confirm that both authors were ultimately referring to Dominicus Gundissalinus's treatise *De divisione philosophiae* for this and others of their definitions of philosophy.

A very similar picture emerges from another introductory text to Porphyry, namely the anonymous prologue to the *Isagoge* commentary known as *Sicut dicit Ysaac*, which Martin Grabmann ascribed to Nicholas of Paris.[18] Claude Lafleur placed the redaction of this text around the year 1250, while Ruedi Imbach has argued that it should be prior to 1246/1247 and that there is no evidence for its attribution to Nicholas of Paris.[19] This prologue commences with a series of general definitions of the nature of philosophy. Among the five definitions that are presented, we find the three definitions just discussed, starting with the etymological definition of philosophy:

> Prima definitio, quam ponit Ysaac, est, quod philosophia est amor sapientiae. Haec est prima definitio large sumendo definitionem ad veram definitionem et ad interpretationem sive etymologiam. Unde dicitur philosophia a *philos* graece, quod est *amor* latine, et *sophos*, quod est *sapientia*, inde philosophia, id est amor sapientiae.
>
> > (The first definition given by Isaac says that philosophy is the love of wisdom. This is the first definition in the large sense of true definition, translation and etymology. For philosophy comes from the Greek *philos*, which means *love* in Latin, and from *sophos*, which means *wisdom*, and from here comes philosophy, which means love of wisdom.)[20]

It is worth noting that here the anonymous author correctly attributes this definition to Isaac Israeli, who is the source for Gundissalinus's *De divisione philosophiae*. Yet, the wording in the prologue to the *Isagoge* commentary does not reflect Isaac Israeli's wording in his *Liber de definitionibus*. The following

---

17 Lafleur, *Quatre Introductions à la philosophie au XIIIe siècle*, pp. 308–09.

18 Grabmann, 'Die logischen Schriften des Nikolaus von Paris', esp. pp. 243–47.

19 Lafleur, *Quatre Introductions à la philosophie au XIIIe siècle*, pp. 138–39. Imbach, 'Einführungen in die Philosophie aus dem XIII. Jahrhundert', pp. 479–80.

20 Imbach, 'Einführungen in die Philosophie aus dem XIII. Jahrhundert', p. 486.

table offers the wording of the definition in the two translations of the *Liber de definitionibus*, attributed to Gerard of Cremona and Gundissalinus respectively:

Trans. Gerard[21]	Trans. Gundissalinus[22]
Descriptio autem eius derivata ex nomine est quod philosophia est amor sapientiae et illud rationatur ex nomine philosophi; philosophus enim compositus est ex *phile* et *sophia*. Et dicitur quidem lingua graeca *phile amator* et *sophia sapientia*. Cumque illud ita sit, tunc manifestum est quod intentio philosophi est amator sapientiae, et cum intentio philosophi sit amator sapientiae, tunc intentio philosophiae est amor sapientiae.	Prima sumpta ex interpretatione nominis est haec, philosophia est amor sapientiae.

Obviously, Gundissalinus's truncated translation of the *Liber de definitionibus* cannot have been the source for the anonymous author of *Sicut dicit Ysaac*; yet, also Gerard's translation, which offers the full development of the definition, is not likely to have served as his *Vorlage*. The compact phrasing of the etymological argument 'unde dicitur philosophia a *philos* graece, quod est *amor* latine, et *sophos*, quod est *sapientia*, inde philosophia, id est amor sapientiae' rather suggests *De divisione philosophiae* as its source: '*Philos* enim graece *amor* dicitur latine; et *sophia* dicitur *sapientia*; inde philosophia est amor sapientiae et philosophus dicitur amator sapientiae'.[23] The fact that *Sicut dicit Ysaac* correctly identifies Isaac Israeli as the author of this definition, but nonetheless quotes the text from Gundissalinus is not surprising, as the same phenomenon occurs in contemporary introductions to philosophy, whose authors try to match Gundissalinus's definitions with their ultimate sources (or what they believe those sources were).[24]

Next, the author of the prologue to the *Isagoge* commentary quotes the fifth of Gundissalinus's definitions from *De divisione philosophiae*, which we have come across before:

> Tertia definitio est: Philosophia est cognitio sui ipsius ab homine, secundum quod dicit Algazel, quod 'cognosce te ipsum et cognosces omnia', quia in homine est substantia at accidens et omne, quod est, aut est substantia aut accidens [...] Haec autem omnia sunt in homine secundum speciem,

21 Isaac Israeli, *Liber de definitionibus*, ed. by Muckle, pp. 302–03.

22 Isaac Israeli, *Liber de definitionibus*, ed. by Muckle, p. 330.

23 Dominicus Gundissalinus, *De divisione philosophiae*, ed. by Fidora and Werner, pp. 58/60. English translation above.

24 The above case of Arnulf of Provence is but one example. See the introduction to Dominicus Gundissalinus, *De divisione philosophiae*, ed. by Fidora and Werner, pp. 45–46.

> quoniam substantiarum alia corporea et corruptibilis, alia incorporea et incorruptibilis. Haec duo sunt in homine. Est enim ibi corpus, quod est substantia corporea et corruptibilis, et est ibi anima, quae est incorporea et incorruptibilis.
>
> > (The third definition is: Philosophy is man's cognition of himself, according to what al-Ġazālī says, that is, 'know yourself and you will know everything', for in man there is substance and accident [...] All these things are in man according to their species, because some substances are corporeal and corruptible, while some are incorporeal and incorruptible. In man there is body which is a corporeal and corruptible substance, and there is soul, which is an incorporeal and incorruptible substance.)[25]

In this instance, the anonymous author does not provide a more precise reference that would disclose the real source used by Gundissalinus. Instead, as before in *Ut ait Tullius*, the definition is erroneously attributed to al-Ġazālī, while Gundissalinus took it, again, from Isaac Israeli's *Liber de definitionibus*.[26] Eventually, *Sicut dicit Ysaac* also credits, for its fifth definition, al-Ġazālī as being the author of definition number three from *De divisione philosophiae*:

> Quinta autem sumpta ab Algazele est, quod philosophia est divinarum humanarumque rerum cognitio.
>
> > (The fifth, which is taken from al-Ġazālī, is that philosophy is the cognition of divine and human things.)[27]

Once more, as in *Ut ait Tullius*, this definition is deemed to derive from Arabic sources, although, as mentioned before, this is the well-known Isidorian definition of philosophy that Gundissalinus includes among his Arabic materials.[28]

I now turn to the subsequent discussion of the various branches and parts of philosophy in *Sicut dicit Ysaac*, which also recalls Gundissalinus's

---

25 Imbach, 'Einführungen in die Philosophie aus dem XIII. Jahrhundert', p. 486.

26 See above note 13.

27 Imbach, 'Einführungen in die Philosophie aus dem XIII. Jahrhundert', p. 487.

28 It is worth noting that the three definitions of philosophy from Gundissalinus's *De divisione philosophiae*, which we have discussed so far, appear in other prologues in thirteenth-century commentary literature as well, such as Robertus Anglicus's commentary on Peter of Spain's *Summulae logicales*. See the fragments edited in de Rijk, 'On the Genuine Text of Peter of Spain's *Summulae logicales*, III', at p. 24: 'Hiis visis, definitiones aliquae notentur. Philosophia multis modis definitur. Primo modo philosophia est amor sapientiae. Alio modo philosophia est sui ipsius cognitio ab homine, id est homo cognoscendo se cognoscit philosophiam. Alio modo philosophia est divinarum et humanarum etc.'. Further prologues to Aristotle commentaries from the first half of the thirteenth century that use these definitions are discussed in Köhler, *Grundlagen des philosophisch-anthropologischen Diskurses im dreizehnten Jahrhundert*, pp. 443–57.

De divisione philosophiae. A very significant example is the tripartition of practical philosophy into ethics, economics, and politics, which *Sicut dicit Ysaac* describes in the following terms:

> Moralis autem dividitur triplex, quia triplex est bonum. Est enim bonum monasticum, et est bonum yconomicum, et est bonum politicum. Bonum monasticum dicitur, quando aliquis debet regere seipsum tantum. [...] Yconomica dicitur ab *yconos*, quod est dispensatio, quando aliquis bene scit dispensare propriae familiae, propriis filiis et filiabus, famulis et famulabus. Politica dicitur a *polis*, quod est civitas, secundum quod huius praelati dicuntur regere civitates ut praepositi et alii. Duae ulitmae partes, scilicet yconomica et politica, sicut quidam dicunt, traduntur in legibus et decretis, alii dicunt, quod traduntur a Tullio in libro *De officiis*. Alii dicunt, quod Aristoteles fecit in lingua arabica quandam scientiam de hoc, quae nobis adhuc non est translata. [...] Monastica autem dividitur secundum duplex est bonum. [...] De primo est *Vetus Ethica*, de secundo bono est *Nova Ethica*.
>
> > (The division of moral philosophy is threefold, because there are three goods. There is a monastic good, an economic good, and a political good. One speaks of a monastic good when somebody has to rule over himself only. [...]. Economics comes from *yconos* which means organization, i.e. when somebody knows well how to organize his household, his sons and daughters, and male and female servants. Politics comes from *polis* which means city, insofar as its principals rule over the cities in their capacity as superiors etc. Concerning the latter two, that is, economy and politics, some say that they are transmitted in laws and decrees, while others say that they are taught by Cicero in his book *On Duty*. Still others say that Aristotle composed a science thereof in Arabic, which has not yet been translated. [...] The monastic discipline, however, is divided according to the two modes of good. [...] The first is dealt with in the *Old Ethics*, the second in the *New Ethics*.)[29]

This tripartite division, which was well known throughout the Middle Ages, can also be found in Gundissalinus's *De divisione philosophiae*:[30]

> De partibus practicae philosophiae. Quarum prima scientia est gubernandi civitatem, quae dicitur politica sive civilis ratio [...]. Secunda est scientia regendi familiam propriam, quae tribus modis regitur, videlicet:

29 Quoted from Grabmann, 'Die logischen Schriften des Nikolaus von Paris', pp. 245–46 (in part also in Imbach, 'Einführungen in die Philosophie aus dem XIII. Jahrhundert', p. 490).

30 On the medieval transmission of a tripartite practical philosophy and its reception and transformation in Gundissalinus's *De divisione philosophiae*, see Fidora, *Die Wissenschaftstheorie des Dominicus Gundissalinus*, pp. 167–79.

> [...] disciplina a vitiis corrigendo, sollicitudo eorum necessitatibus subveniendo, doctrina vero eos honestis artibus instruendo [...]. Tertia est gubernatio sui ipsius. Cui tria sunt necessaria, scilicet: fuga vitiorum, exercitium virtutum et exempla meliorum.
>
> (On the parts of practical philosophy: The first of these is the science of ruling the city, which is called politics or civil reason [...]. The second is the science of ruling one's household, which is done in three ways: [...] discipline corrects the vices; care addresses the needs, and teaching introduces to the practice of the honest arts [...]. The third is about ruling over oneself, wherefore three things are required: avoiding the vices, exercising the virtues, and providing examples of better people.)[31]

A distinctive feature of Gundissalinus's account, in contrast to other twelfth-century descriptions of the parts of practical philosophy, is his definition of all three parts of practical philosophy by means of the kind of direction or rule that each is concerned with, that is, over the city (politics), over the household (economics) and over the individual himself or herself (ethics).[32] Gundissalinus's tripartition and its characteristic description of the specific relations of dominion resonates forcefully with *Sicut dicit Ysaac*. Moreover, it is likely that Gundissalinus is also at the origin of the tantalizing reference in *Sicut dicit Ysaac* to a book by Aristotle on politics, supposedly written in Arabic, though not yet translated. While the information is vague, it can be considered a very early reference to Aristotle's *Politics*.[33] Apparently, the author knew the *Ethica vetus* and the *Ethica nova*, but not Grosseteste's translation of the *Nicomachean Ethics* from 1246/1247, and even less the translation of the *Politics* by William of Moerbeke from *c.* 1260. The fact that he refers to an *Arabic*

---

31 Dominicus Gundissalinus, *De divisione philosophiae*, ed. by Fidora and Werner, pp. 252, 260 and 262.

32 For this important point, see Bertelloni, 'Presupuestos de la recepción de la *Politica* de Aristóteles', p. 47.

33 The threefold division of practical philosophy also appears in other contemporary texts by the Parisian masters of arts, yet without ever mentioning Aristotle's *Politics*; see for instance the prologue to Robertus Anglicus's commentary on the *Summulae logicales*: 'Practica dividitur in monasticam, yoconomicam et politicam. Monastica est quae ordinat hominem ad cognitionem sui; et dicitur a monos, quod est unum, et ycos, quod est custodia; et haec debetur monachis et heremitis et haec datur ab Aristotele in *Ethicis*. Yconomica dicitur quae docet propriam familiam ordinare. Unde yconomus graece idem est quod dispensator latine. Vel dicitur ab Yconomo, primo inventore huius scientiae; et haec a Tullio libro suo *De officiis* enodatur. Politica nominatur quae docet regere populos et civitates. Unde dicitur a polis, quod est civitas, vel pluralitas; et haec traditur in legibus et decretis' (de Rijk, 'On the Genuine Text of Peter of Spain's *Summulae logicales*, III', p. 23). An exception is the *Isagoge* commentary *Secundum quod testatur Ysaac*, which is very close to *Sicut dicit Ysaac*, though not identical with it, as Ruedi Imbach has shown; see the text in Imbach, 'Einführungen in die Philosophie aus dem XIII. Jahrhundert', p. 490, which refers to a work of Aristotle on politics that has not yet been translated.

book on politics by Aristotle hints at the possible origin of his information, namely a text imbued with the Arabic tradition such as Gundissalinus's *De divisione philosophiae*. Indeed, *De divisione philosophiae* yields one of the first explicit Latin testimonies to the existence of Aristotle's *Politics*. Thus, in the chapter on politics in *De divisione philosophiae*, Gundissalinus writes: 'Et haec quidem scientia continetur in libro Aristotelis, qui *Politica* dicitur, et est pars *Ethicae*' (And this science is contained in a book by Aristotle which is called *Politics*, being a part of the *Ethics*).[34] As Cary J. Nederman has emphasized:

> Gundisalvi's reference to the *Politics* suggests that his treatment of political science as distinct from and superior to ethics and economics represented an attempt to reproduce what he regarded as Aristotle's own position. With Gundisalvi, one might say, the 'underground tradition' of Aristotelian political science begins to surface, to acknowledge explicitly its debt to Aristotle.[35]

It is likely therefore that the Aristotelian tripartition of practical philosophy and in particular the precocious reference to Aristotle's *Politics* in *Sicut dicit Ysaac* are based on the account provided by Gundissalinus's *De divisione philosophiae*.[36]

One can also observe the influence of Gundissalinus with regard to the internal divisions of theoretical philosophy and of logic. Thus, *Sicut dicit Ysaac* follows *De divisione philosophiae* in defining the various parts of logic and natural philosophy according to the respective books of the *Corpus aristotelicum*. More specifically, elaborating on al-Fārābī, Gundissalinus claimed that the first part of natural philosophy examines natural bodies with a view to what is common to all of them, and corresponds to Aristotle's *Physics*, the second part of natural philosophy, which deals with simple bodies, to *De caelo*, and the third part, which enquires into the mixture and corruption of the natural body, to *De generatione et corruptione*. The fourth part of natural philosophy deals with the effects and passions of the elements, which are treated in the first three books of *De impressionibus superiorum*, that is, the *Meteorologica*. The last book of Aristotle's *Meteorologica* constitutes the fifth part of natural philosophy, which is concerned with bodies, insofar as they are composed of elements. The sixth part of natural philosophy deals with minerals and metals and is said to be covered by a tract *De mineris*, which belongs to the pseudo-Aristotelian tradition. The seventh part of natural philosophy is dedicated to plants and is presented in yet another pseudo-Aristotelian work, that is, *De vegetabilibus*. The eighth and last part of natural philosophy,

34 Dominicus Gundissalinus, *De divisione philosophiae*, ed. by Fidora and Werner, p. 136.

35 Nederman, 'Aristotelianism and the Origins of "Political Science" in the 12th Century', p. 190.

36 For a more detailed analysis of Gundissalinus's concept of practical philosophy as compared to the interpretation of practical philosophy in the works of the Parisian masters of arts during the thirteenth century (including *Sicut dicit Ysaac*), see Fidora, 'Politik, Religion und Philosophie in den Wissenschaftseinteilungen der Artisten'.

which is identified with the Aristotelian works *De animalibus, De anima,* and *De naturalibus,* i.e. the *Parva naturalia,* is concerned with animate bodies.[37]

Gundissalinus's division of natural philosophy, as well as of logic, in terms of the *Corpus aristotelicum* would turn into the standard account of these disciplines.[38] Also the prologue to *Sicut dicit Ysaac* follows this 'bookish' principle in dividing natural philosophy, as well as logic.[39] The first distinction it draws is between corruptible and incorruptible mobile bodies, the latter of which constitute the domain of enquiry of *De caelo*. Corruptible mobile bodies, in turn, are studied in a general way in the book of *Physics* and in a number of more specific ways in the following ones: *De generatione et corruptione* studies the simple mobile corruptible body; the compound mobile corruptible body is the subject of the *Meteora* when it is inanimate; when animate its consideration falls under *De plantis, De morte et vita, De animalibus, De sensu et sensato, De somnio et vigilia, De memoria et reminiscentia, De anima,* etc., according to further specifications.[40] While the prologue to *Sicut dicit Ysaac* does not reproduce the same sequence and description of subject matters that Gundissalinus laid out, it does obey to the fundamental epistemological intuition of structuring natural philosophy in conformity with the *libri naturales* — an intuition which finds support in *Meteorologica* (I. 1. 338a 20–339a 8), where Aristotle draws a very similar picture.[41]

Although from a philological point of view it is difficult to determine the exact channels through which Gundissalinus's doctrines reached the commentators of the *Isagoge* — either directly or through contemporary

37 See Dominicus Gundissalinus, *De divisione philosophiae,* ed. by Fidora and Werner, pp. 76/78/80 and, for the division of logic pp. 156/58. Cf. also Fidora, 'Aristotelische Wissenschaft als Netzwerk von Wissenschaften'.

38 See the very early commentary on Aristotle's *Physics* (before 1238) by the Franciscan philosopher and theologian Richard Rufus: Richard Rufus, *In Physicam Aristotelis,* ed. by Wood, p. 93 (natural philosophy). See also the introductions to philosophy from the first half of the thirteenth century, e.g. in the anonymous *Philosophica disciplina* or Arnulf of Provence's *Divisio scientiarum* in Lafleur, *Quatre Introductions à la philosophie au* XIII*e siècle*: *Philosophica disciplina,* pp. 262–64 (natural philosophy) and pp. 282–83 (logic), *Divisio scientiarum,* pp. 332–33 (natural philosophy) and pp. 343–44 (logic).

39 For the subdivisions of logic, see Grabmann, 'Die logischen Schriften des Nikolaus von Paris', p. 247.

40 See the text in Imbach, 'Einführungen in die Philosophie aus dem XIII. Jahrhundert', pp. 492–93.

41 Even though in a far less developped form, the division of natural philosophy according to the *libri naturales* also appears in Robertus Anglicus's commentary on the *Summulae logicales*: 'Scientia naturalis dividitur secundum divisionem sui subiecti. In naturali philosophia subiectum est corpus mobile. Unde natura ab Aristotele dicitur principium motus et quietis. Aut agit de corpore mobili non contracto ad aliquam materiam; et sic liber *Physicorum,* qui Canon naturalis Averroys nuncupatur. Si fit de corpore mobili contracto, aut contrahitur ad suum causatum; et sic est *Liber caeli et mundi* in quo determinatur de motibus corporum supracaelestium, quae incorruptibilia sunt' (de Rijk, 'On the Genuine Text of Peter of Spain's *Summulae logicales,* III', p. 23).

introductions to philosophy — there can be no doubt that *Ut ait Tullius* and *Sicut dicit Ysaac* incorporate characteristic features of Gundissalinus's division of philosophy into their epistemological preliminaries. Gundissalinus's definitions of philosophy and the internal divisions of both practical and theoretical philosophy were thus essential elements of these prologues to philosophical commentaries, which offered brief sketches of the whole of philosophy and its parts.

Peter of Ireland and his Commentary on Aristotle's *Peri hermeneias*

The Neapolitan master Peter of Ireland and his commentary on Aristotle's *Peri hermeneias* deserve particular attention. Slightly later than the two texts on Porphyry's *Isagoge* and quite independent of the introductions to philosophy, this work opens with a prologue that is based completely on Gundissalinus's *De divisione philosophiae*.[42] The editor of the text, Michael Dunne, has suggested dividing this prologue into four parts: 1) a *definitio philosophiae*, 2) a *divisio philosophiae*, 3) a section *De logica*, and 4) a section *De intellectu*. For sections 1) to 3), Dunne has identified the sources in Gundissalinus's prologue to *De divisione philosophiae*, and, as I shall show, the section *De intellectu* also draws on this work, namely on its chapter on logic.

1) The first section consists of a single definition, which occupies the opening line of Peter of Ireland's prologue. Accordingly, philosophy is defined as the knowledge of everything that human understanding can possibly attain: 'philosophia est cognitio veritatis omnium rerum et secundum quod possibile est homini comprehendere'.[43] This text closely resembles a passage in Gundissalinus's prologue to *De divisione philosophiae*, where, however, it is not presented as a definition of philosophy *sensu stricto*, but rather as an explanation of its goal, that is:

> Intentio philosophiae est comprehendere veritatem omnium, quae sunt, quantum possibile est homini.
>
> > (The intention of philosophy is to comprehend the truth of all things there are, insofar as it is possible for man to do so.)[44]

---

42 The dating of the work is controversial, oscillating between 1250 and 1260–1268. See Dalgaard, 'Peter of Ireland's Commentary on Aristotle's *Peri hermeneias*', p. 9, and Michael Dunne in his introduction to Peter of Ireland, *Expositio et quaestiones in librum Aristotelis Peryermenias seu De interpretatione*, ed. by Dunne, pp. xv–xvi. In my opinion the text was written during the 1250s or the very beginning of the 1260s.

43 Peter of Ireland, *Expositio et quaestiones in librum Aristotelis Peryermenias*, ed. by Dunne, p. 3.

44 Dominicus Gundissalinus, *De divisione philosophiae*, ed. by Fidora and Werner, p. 60; English translation from Grant, *A Source Book in Medieval Science*, p. 60.

Gundissalinus took this from Avicenna's *Logica* from the *Kitāb al-shifā'* which was translated in Toledo by Avendauth, with whom Gundissalinus worked closely:[45]

> Intentio philosophiae est comprehendere veritatem omnium rerum, quantum possibile est homini comprehendere.
>
> (The intention of philosophy is to comprehend the truth of all things, insofar as it is possible for man to comprehend.)[46]

2) The second section, which elaborates on the *divisio philosophiae*, specifies that this 'knowledge of everything that human understanding can possibly attain' falls into the knowledge of things that are the result of human action, and those that are not:

> Unde quaedam pars philosophiae est contemplatio veritatis omnium rerum quae non sunt ex opere nostro, et ista dicitur theorica; altera est contemplatio omnium rerum quae sunt ex opere nostro, et ista dicitur practica.
>
> (Therefore, one part of philosophy is concerned with the contemplation of the truth of all things that are not from our own work — and this part is called theoretical —, while the other is concerned with the contemplation of all things that are from our own work — and this part is called practical.)[47]

The same fundamental distinction between objects of study that are under our control and those that are not, appears in Gundissalinus's prologue to *De divisione philosophiae*:

> Sed omnium, quae sunt, alia sunt ex nostro opere et nostra voluntate, ut nostra humana opera sicut leges, constitutiones, Dei cultus exercitia, bella et alia huiusmodi; alia sunt non ex nostro opere nec ex nostra voluntate, ut Deus, angeli, caelum, terra, vegetabilia, animalia, metalla, spiritus et omnia naturalia.
>
> (But of all things that are, some are from our own work and will — our human works, such as laws, constitutions, religious exercises, wars and other things of this kind. Others are not from our own work or will — such as God, angels, heaven, earth, vegetables, animals, metals, spirits and all natural things.)[48]

And as Peter of Ireland does, Gundissalinus goes on to call the two parts of philosophy *practica* and *theorica*:

---

45 On Gundissalinus and Avendauth see, for instance, Fidora, 'Religious Diversity and the Philosophical Translations of 12th-Century Toledo'.

46 Avicenna, *Logica*, 1, fol. 2ra.

47 Peter of Ireland, *Expositio et quaestiones in librum Aristotelis Peryermenias*, ed. by Dunne, p. 3.

48 Dominicus Gundissalinus, *De divisione philosophiae*, ed. by Fidora and Werner, pp. 60/62; English translation from Grant, *A Source Book in Medieval Science*, p. 60.

> Una enim est pars philosophiae, quae facit cognoscere, quid debeat agi, et haec dicitur practica; et altera est, quae facit cognoscere, quid debeat intelligi, et haec est theorica. [...] Nam finis speculativae est conceptio sententiae ad intelligendum; finis vero practicae est conceptio sententiae ad agendum. Partes igitur, in quas primum philosophia dividitur, hae sunt: scilicet theorica et practica.
>
> > (For there is one part of philosophy which makes us know what ought to be done and this is called practical; and there is another one which makes us know what ought to be understood and this is theoretical. [...] The goal of speculative philosophy is to acquire a rule for understanding, while the goal of practical philosophy is to acquire a rule for acting. The parts, therefore, into which philosophy is first divided are theoretical and practical.)[49]

In this, Gundissalinus again follows Avicenna's *Logica*, though he introduces some terminological changes that Peter of Ireland's wording replicates. For instance, Avicenna's *Logica* speaks of *philosophia activa* and *speculativa*,[50] whereas Gundissalinus and Peter have *theorica* and *practica*.

Gundissalinus and Avicenna are also the basis for Peter's further development of theoretical philosophy. Thus, Peter of Ireland establishes the following subdivision of theoretical philosophy:

> Res autem quae sunt praeter operationem nostram sunt dupliciter: aut enim sunt possibiles permisceri materiae, aut non sunt possibiles, sicut Deus et angeli. Possibiles permisceri sunt aut quod impossibile est eas aliter esse vel intelligi nulla permiscione, prout sunt humanitas et asinitas et similia; aut possibiles sunt non permisceri secundum esse, quamvis accidat eis permisceri, sicut sunt unitas et multitudo, causa vel causata [...].
>
> > (The things, however, which do not depend on our operation are twofold: either it is possible for them to be mixed with matter or not, as is the case of God and the angels. Regarding those which can be mixed with matter, either it is not possible for them to exist otherwise and to be thought of without any mixture, like being a man or a donkey; or it is possible for them to exist without mixture, even though such mixture may happen to them, e.g. unity, multiplicity, cause and effects.)[51]

For this passage, one finds in Gundissalinus:

> Ea vero, quae sunt non ex nostro opere nec ex nostra voluntate, secundum quosdam ex philosophis primo dividuntur in duo; nam eorum alia sunt,

---

49 Gundissalinus, *De divisione philosophiae*, ed. by Fidora and Werner, p. 64. English translation from Grant, *A Source Book in Medieval Science*, p. 61 (slightly modified).

50 See Avicenna, *Logica*, 1, fol. 2ra.

51 Peter of Ireland, *Expositio et quaestiones in librum Aristotelis Peryermenias*, ed. by Dunne, p. 3.

> quibus nullus unquam accidit motus, ut Deus et angelus; et alia sunt, quibus accidit motus, ut identitas, unitas, multitudo et causalitas. Sed haec, quibus accidit motus, similiter dividuntur in duo; nam eorum alia sunt, quae numquam possunt esse sine motu, ut humanitas et quadratura; et alia sunt, quae possunt esse sine motu, ut unitas et causa.
>
> > (According to some philosophers, the things which are not from our work and will are first divided into two. For some of these are never related to motion, such as God and angel, while others are sometimes related to motion, such as identity, unity, multiplicity and causality. In turn, those related to motion are likewise divided into two, as some of them can never be without motion, such as being a man or a square, while others can be without motion, such as unity and cause.)[52]

Although Peter replaces the criterion of motion (*motus*) with matter-relatedness (*materia*), which Gundissalinus and Avicenna introduce into their discourses at a later point, he follows the general structure they provide as well as their examples. With regard to the latter, it is worth noting that Peter speaks of 'asinitas', where Gundissalinus uses the example of 'quadratura'. In this, Peter sides with Avicenna who has 'forma asinina';[53] Peter was probably using both texts simultaneously, that is, Gundissalinus's *De divisione philosophiae* and Avicenna's *Logica*.

Before turning to practical philosophy, Peter concludes his remarks on theoretical philosophy with a reference to its famous division into metaphysics, mathematics, and physics according to Aristotle's *Physics*. II. 7. 198a 29–31:

> Et hoc est quod habetur in II *Physicae* quod triplex est negotium, unum scilicet de immobilibus, alterum de mobilibus et incorruptibilibus, tertium de mobilibus corruptibilibus.
>
> > (And this is what is said in the second book of the *Physics*, namely that the inquiry is threefold: one inquiry deals with what is immobile, another one with what is mobile and incorruptible, and a third one with what is mobile and corruptible.)[54]

It is worth noting that a similar reference to Aristotle's *Physics* is placed at the end of the discussion of the three parts of theoretical philosophy in Gundissalinus's prologue to *De divisione philosophiae*, immediately before introducing the division of practical philosophy:

> Unde Aristoteles: Ideo scientiarum sunt species tres, quoniam una speculatur, quod movetur et corrumpitur ut naturalis; et secunda, quod

---

52 Dominicus Gundissalinus, *De divisione philosophiae*, ed. by Fidora and Werner, pp. 64/66.

53 See Avicenna, *Logica*, 1, fol. 2ra.

54 Peter of Ireland, *Expositio et quaestiones in librum Aristotelis Peryermenias*, ed. by Dunne, p. 4.

movetur et non corrumpitur ut disciplinalis; tertia considerat, quod nec movetur nec corrumpitur ut divina.

> (Whence Aristotle says: Therefore there are three species of sciences, since one investigates what is moved and corruptible, i.e. natural science; the second what is moved and incorruptible, i.e. mathematics; and the third considers what is neither moved nor corruptible, i.e. divine science.)[55]

While there can be no doubt that Peter of Ireland quotes directly from Aristotle and not through Gundissalinus (who translated from Arabic),[56] it is still remarkable that he introduces this very pertinent quotation at the same place as it appears in Gundissalinus's argument. In fact, a contemporary of Peter, William of Melitona, also read Gundissalinus together with Aristotle on the threefold division of theoretical philosophy.[57]

3) After defining philosophy and dividing it into a practical and a theoretical branch with its respective parts, the section *De logica* raises the central question of whether this description must be considered exhaustive, and if so, what would be the place of logic, which falls neither under theoretical nor under practical philosophy.[58] Peter's answer claims that the relation of logic to philosophy is twofold. On the one hand, it is a part of philosophy, insofar as things can be considered either in themselves, which is how they are studied in practical and theoretical philosophy, or insofar as they are in the soul, where they have particular accidents such as 'subiectio', 'praedicatio', 'substantialitas', 'accidentalitas' and 'proportionabilitas'. These specific accidents are the object of study of a specific part of philosophy, namely logic. On the other hand, logic is not only a part of philosophy, but also its instrument, insofar as its general function is to provide the rules by which one passes from what is not known to what is known. Logic is therefore both an instrument and a part of philosophy. This solution resumes Gundissalinus's characterization of the twofold nature of logic in his prologue to *De divisione philosophiae*:

Logica vero, secundum quod utilis est ad veritatem in se et in aliis scientiis inveniendam, instrumentum, sed secundum quod philosophia subiecti eius dispositiones inquirit, sicut et de ceteris, pars eius est.

> (Logic, however, is in one respect an instrument useful for finding out the truth in itself and in the other sciences, and in another respect

55 Dominicus Gundissalinus, *De divisione philosophiae*, ed. by Fidora and Werner, pp. 68/70; English translation from Grant, *A Source Book in Medieval Science*, pp. 61–62 (slightly modified).

56 See Fidora, 'Nota sobre Domingo Gundisalvo y el *Aristoteles arabus*'.

57 William of Melitona, *Quaestiones de sacramentis*, ed. by Piana and Gál, vol. 2, p. 619: 'Ut determinatur a diversis philosophis, scilicet Aristotele et Toletano, res dividuntur, et secundum rerum divisiones consequenter dividuntur scientiae'.

58 Cf. Peter of Ireland, *Expositio et quaestiones in librum Aristotelis Peryermenias*, ed. by Dunne, pp. 4–5.

> it is a part of philosophy according as philosophy investigates the dispositions of its subject (just as it investigates other dispositions.)[59]

4) The fourth section, which Michael Dunne has labelled 'De intellectu', offers a brief gnoseological account that explains how logic, as an instrument of philosophy, leads from what is not known to what is known. This account is based on the two key notions of 'intellectus' and 'credulitas'. The latter of these betrays its Arabic source, namely al-Ġazālī's famous doctrine of *taṣawwur* (concept or representation) and *taṣdīq* (judgement), which Peter of Ireland most probably came across in the chapter on logic in Gundissalinus's *De divisione philosophiae*:

Peter of Ireland, *In Peryermenias*[60]	**Gundissalinus,** *De divis. phil.*[61]
Cognitum enim potest intelligi dupliciter: aut dicitur cognitum cuius intellectus est cognitus, sicut cum intentio intellectiva nominis apprehenditur per nomen sine vero et falso, ut intellectus huius nominis 'homo', aut intentio intellectiva per orationem sine vero et falso, ut intellectus huius 'fac ignem'; alio modo dicitur cognitum quando cum intellectu haberi credulitas de re; et dicitur haberi credulitas cum intellectu haberi consensus de esse rei, ut 'homo est albus'.	Duobus autem modis cognoscitur res, imaginatione scilicet et credulitate: imaginatione, ut ex significatione sui nominis, ut cum dictur 'homo' vel 'animal'; credulitate, ut cum dicitur 'homo est animal'. Imaginatio autem necessario praecedit credulitatem. Nisi enim prius cognoveris significationem huius nominis 'homo' vel 'animal', non credes hoc, quod dicitur 'homo est animal'. Imaginatio autem aliquando recipitur sine probatione.

While Peter replaces the translation of *taṣawwur* as 'imaginatio' with 'intellectus', his argument follows the lines of Gundissalinus's presentation of the two terms in question.[62]

As we have seen, all four sections of the prologue to Peter of Ireland's commentary on Aristotle's *Peri hermeneias* rely heavily on Gundissalinus's treatise *De divisione philosophiae*. Thus, although Gundissalinus himself was drawing on various sources for his account, such as Avicenna or al-Ġazālī, which Peter was certainly familiar with, the very organization of these materials

---

59 Dominicus Gundissalinus, *De divisione philosophiae*, ed. by Fidora and Werner, p. 74; English translation from Grant, *A Source Book in Medieval Science*, p. 62. See also the development of these remarks in the chapter on logic in *De divisione philosophiae*, p. 150. Michael B. Crowe has pointed to an alternative source, namely Boethius's commentary on the *Isagoge*, which is quoted on this point by Peter Abaelard in his Glosses on Porphyry. However, the argument in Peter of Ireland is much closer to Gundissalinus than to Boethius (not to mention Abelard). See Crowe, 'Peter of Ireland's Approach to Metaphysics', p. 158.

60 Peter of Ireland, *Expositio et quaestiones in librum Aristotelis Peryermenias*, ed. by Dunne, p. 5.

61 Dominicus Gundissalinus, *De divisione philosophiae*, ed. by Fidora and Werner, p. 164.

62 Peter's wording and examples are in fact closer to Gundissalinus's treatise than to its source, i.e. al-Ġazālī's *Logica* from the *Maqāṣid al-falāsifa*. Cf. al-Ġazālī, *Logica*, ed. by Lohr, p. 239.

corresponds so closely to Gundissalinus's treatise that it can be considered Peter's direct model.[63]

Through Peter's commentary, Gundissalinus's division of philosophy probably had an impact on Thomas Aquinas and other philosophers and theologians from the second half of the thirteenth century. For a long time, it was believed that Peter of Ireland taught at the University of Naples when Thomas Aquinas studied there in his early years (1239–1244).[64] While doubt has been cast on Peter of Ireland's role as a teacher of Thomas Aquinas, the question remains under discussion.[65] However, whether or not a direct pupil of Peter, later on, Thomas seems to have used the former's commentary on *Peri hermeneias* for his own exposition of the work.[66]

The *divisio philosophiae* in Thomas Aquinas's Aristotle Commentaries

Thomas Aquinas's commentaries on Aristotle's oeuvre are clearly part of the tradition that has been reconstructed so far; many of his very influential expositions open with concise epistemological prologues. Analysing Aquinas's prologues, Francis Cheneval and Ruedi Imbach have pointed out, among others things, the strict logical order that Aquinas establishes in these texts among the different books of the *Corpus aristotelicum*.[67] Thus, in a way that greatly resembles Gundissalinus and al-Fārābī, Aquinas arranges both logic and natural philosophy along the books of the Aristotelian corpus. Logic, for instance, is divided in his commentary on *Peri hermeneias* as follows:

> Cum autem logica dicatur rationalis scientia, necesse est quod eius consideratio versetur circa ea quae pertinent ad tres praedictas operationes rationis; de hiis igitur quae pertinent ad primam operationem intellectus, id est de hiis quae simplici intellectu concipiuntur, determinat Aristoteles in libro *Praedicamentorum*; de hiis vero quae pertinent ad secundam operationem, scilicet de enuntiatione affirmativa et negativa, determinat Philosophus in libro *Peryermenias*; de hiis vero quae pertinent ad tertiam operationem

---

63 Lidia Lanza has recently suggested that Peter might have had access to Gundissalinus's doctrines through Michael Scot's *Divisio philosophiae*. Yet, while being strongly dependent on Gundissalinus, Michael's division does not include the texts adduced here for comparison. See Lanza, 'Artisten', p. 997.

64 See Grabmann, 'Magister Petrus von Hibernia. Der Jugendlehrer des heil. Thomas von Aquin'.

65 See Robiglio, '"Neapolitan Gold": A Note on William of Tocco and Peter of Ireland', the reply by Dunne, 'Concerning "Neapolitan Gold"', and further.

66 See René-Antoine Gauthier's introduction to Ed. Leonina 1*/1, pp. 68* and 86.

67 Francis Cheneval and Ruedi Imbach have underscored the epistemological function of Aquinas's prologues to his commentaries in their Latin-German edition: Thomas Aquinas, *Prologe zu den Aristoteles-Kommentaren*, pp. lxvi–lxviii.

determinat Aristoteles in libro *Priorum* et in consequentibus, in quibus agitur de syllogismo simpliciter et de diversis syllogismorum et argumentationum speciebus, quibus ratio de uno procedit ad aliud; et ideo secundum praedictum ordinem trium operationum liber *Praedicamentorum* ad librum *Peryermenias* ordinatur, qui ordinatur ad librum *Priorum* et consequentes.

> (Since logic is called rational science, it must direct its consideration to the things that belong to the three operations of reason we have mentioned. Accordingly, Aristotle treats those belonging to the first operation of the intellect, that is, what is grasped by the simple concept, in the book *Praedicamentorum*; those belonging to the second operation, i.e., affirmative and negative propositions, in the book *Perihermeneias*; those belonging to the third operation in the book *Priora* and the books following it in which he treats the syllogism absolutely, the different kinds of syllogism, and the species of argumentation by which reason proceeds from one thing to another. And since the three operations of reason are ordered to each other so are the books: the *Praedicamenta* to the *Perihermeneias* and the *Perihermeneias* to the *Priora* and the books following it.)[68]

The distinction between conceptual logic — identified with the *Categories* — propositional logic — identified with *De interpretatione* — and syllogistic logic — identified with the *Analytica priora* — and their orientation towards demonstrative logic, that is, the *Analytica posteriora*, clearly echoes *De divisione philosophiae*, and would in fact remain authoritative until the *Logique de Port Royal*.[69]

No less important in Aquinas's prologues is the division of natural philosophy in terms of Aristotle's *libri naturales*. This division appears in the prologues to the commentaries on Aristotle's *Meteorologica* and his *De sensu et sensato*, and, most prominently, in that to his commentary on the *Physics*:

> Sed quia ea quae consequuntur aliquod commune, prius et seorsum determinanda sunt, ne oporteat ea multoties pertractando omnes partes illius communis repetere, necessarium fuit quod praemitteretur in scientia naturali unus liber, in quo tractaretur de iis [...] Hic autem est liber *Physicorum* [...] Sequuntur autem ad hunc librum alii libri scientiae naturalis, in quibus tractatur de speciebus mobilium: puta in libro *De caelo* de mobili secundum motum localem, qui est prima species motus; in libro autem *De generatione*, de motu ad formam et primis mobilibus, scilicet elementis, quantum ad transmutationes eorum in communi; quantum vero ad speciales eorum transmutationes, in libro *Meteororum*;

68 Thomas Aquinas, *Expositio libri Peri hermeneias*, Ed. Leonina 1*/1, p. 5. English translation from Thomas Aquinas, *Aristotle on Interpretation*, trans. by Oesterle, p. 17 (slightly modified).

69 Cf. Dominicus Gundissalinus, *De divisione philosophiae*, ed. by Fidora and Werner, pp. 156/58, and Höffe, *Aristoteles*, p. 38.

de mobilibus vero mixtis inanimatis, in libro *De mineralibus*; de animatis vero, in libro *De anima* et consequentibus ad ipsum.

> (Those things which are consequent upon something common must be treated first and separately. Otherwise it becomes necessary to repeat such things many times while discussing each instance of that which is common. Therefore it was necessary that one book in natural science be set forth in which those things which are consequent upon mobile being in common are treated [...]. This, then, is the book, *The Physics* [...] Moreover, after *The Physics* there are other books of natural science in which the species of motion are treated. Thus in the *De caelo* we treat the mobile according to local motion, which is the first species of motion. In the *De generatione*, we treat of motion to form and of the first mobile things, i.e., the elements, with respect to the common aspects of their changes. Their special changes are considered in the book *Meteororum*. In the book *De mineralibus*, we consider the mobile mixed bodies which are non-living. Living bodies are considered in the book *De anima* and the books which follow it.)[70]

Here again, one comes across the division of natural philosophy as devised by Gundissalinus and al-Fārābī, which closely follows Aristotle's natural books and their subjects, namely mobile being and the different kinds of bodies.[71]

Another fundamental distinction that is reminiscent of Gundissalinus is the dichotomy of practical versus theoretical philosophy (plus logic) based on whether the object of study belongs to the realm of human *opera(tio)* and *voluntas* or not, which Aquinas refers to in the prologue to his commentary on the *Nicomachean Ethics*.[72]

While these and other reflections in Aquinas's prologues need not necessarily go back to Gundissalinus in a direct manner, they embed his commentaries in a decidedly epistemological tradition of reading and interpreting Aristotle's works, for which Gundissalinus's approach had set the tone.

---

70 Thomas Aquinas, *In Aristotelis libros Physicorum*, Ed. Leonina II, p. 4. English translation from Thomas Aquinas, *Commentary on Aristotle's Physics*, trans. by Blackwell, Spath and Thirlkel, p. 2.

71 Cf. Dominicus Gundissalinus, *De divisione philosophiae*, ed. by Fidora and Werner, pp. 76/78/80 (summarized above).

72 Thomas Aquinas, *Sententia libri Ethicorum*, Ed. Leonina XL. II/1, p. 4: 'Et quia consideratio rationis per habitum scientiae perficitur, secundum hos diversos ordines quos proprie ratio considerat sunt diversae scientiae: nam ad philosophiam naturalem pertinet considerare ordinem rerum quem ratio humana considerat sed non facit, ita quod sub naturali philosophia comprehendamus et mathematicam et metaphysicam; ordo autem quem ratio considerando facit in proprio actu pertinet ad rationalem philosophiam, cuius est considerare ordinem partium orationis ad invicem et ordinem principiorum in conclusiones; ordo autem actionum voluntariarum pertinet ad considerationem moralis philosophiae; ordo autem quem ratio considerando facit in rebus exterioribus constitutis per rationem humanam pertinet ad artes mechanicas'. Cf. Dominicus Gundissalinus, *De divisione philosophiae*, ed. by Fidora and Werner, pp. 60/62.

Conclusion

As a final example of Gundissalinus's sustained influence I would like briefly to point to the fourteenth-century philosopher John of Jandun. Similar to Thomas Aquinas, John starts out the prologue to his *Questions on Aristotle's Physics* with the division of natural philosophy according to the *libri naturales*:

> Philosophia naturalis habet sex partes principales: Prima considerat de ente mobile [...] in communi; et haec traditur in octo libris *Physicorum*. Secunda pars est de ente mobile [...], in qua considerantur corpora simplicia [...] et haec traditur in libro *Caeli et mundi*. Tertia pars est de ente mobile [...], in qua consideratur de generatione et corruptione in communi [...]. Quarta vero pars principalis [...] considerat de ente mobile in ordine ad quasdam passiones [...] et haec traditur in libro *Meteororum*. [...] Quinta pars principalis est de ente mobile mixto inanimato, scilicet de lapidibus et metallis [...] sed illam partem non habemus ab Aristotele. [...] Sexta pars est de corporibus animatis [...].
>
> > (Natural philosophy has six main parts: The first considers mobile being [...] in general; and it is contained in the eight books of the *Physics*. The second part deals with mobile being [...] considering simple bodies [...] and it is contained in *De caelo*. The third part is about mobile being [...] considering generation and corruption in general [...]. The fourth principal part [...] considers mobile being with regard to certain passions [...] and it is contained in the *Meteorologica*. [...] The fifth principal part is about mobile mixed inanimate being, like stones and metals [...], but this part has not come down to us from Aristotle [...] The sixth part is about animate bodies [...] [In what follows, John establishes several subdivisions indicating the relevant Aristotelian book for each of them].)[73]

More than the division of natural philosophy in itself, which we have come across several times already, what is interesting here is that John of Jandun explicitly places it in the original context of his epistemological sources by accompanying it with two definitions of philosophy, namely:

> Philosophia est universi causae cognitio, ut dicit Algazel in tractatu *De divisione philosophiae*. Et Robertus in sui tractatu *De ortu scientiarum* dicit quod philosophia est humanarum divinarum rerumque (*sic*!) cognitio cum studio bene vivendi coniuncta.
>
> > (Philosophy is the knowledge of the cause of everything, as al-Ġazālī says in *De divisione philosophiae*. And Robert says in his treatise *De*

---

73 John of Jandun, *Quaestiones in libros Physicorum Aristotelis*, prooemium (no pagination).

ortu scientiarum that philosophy is the cognition of human and divine things joined with a zeal for living well.)[74]

The first of these definitions, which also occurs in the introductions to philosophy and in *Ut ait Tullius*, could be related to Gundissalinus's above fifth definition, though it is more likely to be an adaptation from al-Ġazālī's *Metaphysica*.[75] The second definition is quoted from Robert Kilwardby's *De ortu scientiarum*,[76] a work that is heavily indebted to Gundissalinus's *De divisione philosophiae*, not only for this definition (which is the third definition above), but also for many of its chapters. With these quotations, which also appear in others of his Aristotle commentaries,[77] John of Jandun acknowledges the historical origin of his epistemological reflections, namely the divisions of philosophy in the tradition of Gundissalinus's encyclopaedia.

The fact that as late as the fourteenth century John of Jandun still alluded to Gundissalinus's definitions and his division of philosophy in his commentaries bespeaks the *longue durée* of these doctrines. Surveying the authors and texts analysed here, it can therefore be said that Gundissalinus's *De divisione philosophiae* not only lay the groundwork for the introductions to philosophy during the first half of the thirteenth century that overtly followed this text as their model. Although less explicit, the influence of Gundissalinus's Arabicized division of philosophy clearly extended further so as to embrace philosophical commentary literature, thus shaping the interpretation of the *Corpus aristotelicum* and other standard texts throughout the thirteenth century and beyond. From this perspective, the study of the reception of Gundissalinus's epistemology is not only of historical interest, but it proves to be a showcase of how philosophical ideas migrate from one academic genre to another, allowing for the continuity of seminal discussions in changing historical and doctrinal contexts.

---

74 John of Jandun, *Quaestiones in libros Physicorum Aristotelis*, prooemium (no pagination).

75 See Dahan, 'Une introduction à l'étude de la philosophie: *Ut ait Tullius*', p. 44 and Lafleur, *Quatre Introductions à la philosophie au* XIII*e siècle*, p. 181. al-Ġazālī, *Algazel's Metaphysics*, ed. by Muckle, p. 4: '[…] ad hoc ut describatur in animabus nostris forma universi esse […]'.

76 Robert Kilwardby, *De ortu scientiarum*, ed. by Judy, p. 10.

77 See Grignaschi, 'Le *De divisione philosophiae* de Dominicus Gundissalinus et les *Quaestiones II–V in Sextum Metaphysicorum* de Jean de Jandun', p. 55.

Bibliography

Manuscripts

Oxford, Bodleian Library, MS Digby 76

Primary Sources

Avicenna, *Logica* (Venezia: Bonetus Locatellus for Octavianus Scotus, 1508, reprinted Frankfurt am Main: Minerva, 1961)

Dahan, Gilbert, 'Une introduction à l'étude de la philosophie: *Ut ait Tullius*', in *L'enseignement de la philosophie au XIIIe siècle. Autour du 'Guide de l'Étudiant' du ms. Ripoll 109*, ed. by Claude Lafleur (Turnhout: Brepols, 1997), pp. 3–58

Dominicus Gundissalinus, *De divisione philosophiae — Über die Einteilung der Philosophie*, ed. and German trans. by Alexander Fidora and Dorothée Werner (Freiburg im Breisgau: Herder, 2007)

al-Ġazālī, *Algazel's Metaphysics: A Mediaeval Translation*, ed. by Joseph Thomas Muckle (Toronto: St Michael's College, 1933)

———, *Logica*, ed. by Charles Lohr, *Traditio*, 21 (1965), 223–90

Isaac Israeli, *Liber de definitionibus*, ed. by Joseph Thomas Muckle, *Archives d'histoire doctrinale et littéraire du Moyen Âge*, 12–13 (1937–1938), 300–28 (= trans. Gerard) and 328–40 (= trans. Gundissalinus)

John of Jandun, *Quaestiones in libros Physicorum Aristotelis* (Venezia: Hieronymus de Sanctis and Johannes Lucilius Santritter, for Petrus Benzon and Petrus de Plasiis, Cremonensis, 1488)

Lafleur, Claude, *Quatre Introductions à la philosophie au XIIIe siècle. Textes critiques et étude historique* (Montréal: Université de Montréal, 1988)

Peter of Ireland, *Expositio et quaestiones in librum Aristotelis Peryermenias seu De interpretatione*, ed. by Michael Dunne (Louvain: Éditions de l'Institut Supérieur de Philosophie, 1996)

Richard Rufus, *In Physicam Aristotelis*, ed. by Rega Wood (Oxford: Oxford University Press, 2003)

Robert Kilwardby, *De ortu scientiarum*, ed. by Albert G. Judy (Oxford: The British Academy, 1976)

Thomas Aquinas, *In Aristotelis libri Physicorum*, cura et sudio Fratrum Praedicatorum, Ed. Leonina II (Roma: Ex Typographia Polyglotta S. C. de Propaganda Fide, 1884)

———, *Aristotle on Interpretation: Commentary by Thomas Aquinas Finished by Cardinal Cajetan*, trans. by Jean T. Oesterle (Milwaukee: Marquette University Press, 1962)

———, *Sententia libri Ethicorum*, cura et sudio Fratrum Praedicatorum, Ed. Leonina XLVII. 1 (Roma: Ad Sanctae Sabinae, 1969)

———, *Expositio libri Peri hermeneias, Editio altera retractata*, cura et sudio Fratrum Praedicatorum, Ed. Leonina I*/1 (Roma: Commissio Leonina, 1989)

———, *Prologe zu den Aristoteles-Kommentaren*, ed. and German trans. by Francis Cheneval and Ruedi Imbach (Frankfurt am Main: Klostermann, 1993)
———, *Commentary on Aristotle's Physics*, trans. by Richard J. Blackwell, Richard J. Spath, and W. Edmund Thirlkel (Notre Dame: Dumb Ox Books, 1999)
William of Melitona, *Quaestiones de sacramentis*, ed. by Caelestinus Piana and Gedeon Gál, 2 vols (Quaracchi: Ex Typographia Collegii S. Bonaventurae, 1961)

Secondary Studies

Bertelloni, Francisco, 'Presupuestos de la recepción de la *Politica* de Aristóteles', in *Aristotelica et lulliana, magistro doctissimo Charles H. Lohr septuagesimum annum feliciter agenti dedicata*, ed. by Fernando Domínguez, Ruedi Imbach, Theodor Pindl, and Peter Walter (Steenbrugge: In Abbatia S. Petri, 1995), pp. 34–54
Chroust, Anton-Hermann, 'The Definitions of Philosophy in the *De divisione philosophiae* of Dominicus Gundissalinus', *New Scholasticism*, 25 (1951), 253–81
Crowe, Michael B., 'Peter of Ireland's Approach to Metaphysics', in *Die Metaphysik im Mittelalter. Ihr Ursprung und ihre Bedeutung*, ed. by Paul Wilpert, Miscellanea Mediaevalia, 2 (Berlin: de Gruyter, 1963), pp. 154–60
Dalgaard, Karen Elisabeth, 'Peter of Ireland's Commentary on Aristotle's *Peri hermeneias*', *Cahiers de l'Institut du Moyen Âge grec et latin*, 43 (1982), 3–44
Dunne, Michael, 'Concerning "Neapolitan Gold": A Note on William of Tocco and Peter of Ireland', *Bulletin de philosophie médiévale*, 45 (2003), 61–66
Fidora, Alexander, 'Nota sobre Domingo Gundisalvo y el *Aristoteles arabus*', *al-Qanṭara*, 23.1 (2002), 201–08
———, *Die Wissenschaftstheorie des Dominicus Gundissalinus — Voraussetzungen und Konsequenzen des zweiten Anfangs der aristotelischen Philosophie im 12. Jahrhundert* (Berlin: Akademie, 2003)
———, 'Politik, Religion und Philosophie in den Wissenschaftseinteilungen der Artisten im 13. Jahrhundert', in *Politischer Aristotelismus und Religion in Mittelalter und Früher Neuzeit*, ed. by Alexander Fidora, Johannes Fried, Matthias Lutz-Bachmann, and Luise Schorn-Schütte (Berlin: Akademie, 2007), pp. 27–36
———, 'Aristotelische Wissenschaft als Netzwerk von Wissenschaften: Die Rezeption der aristotelischen Wissenschaftstheorie bei al-Fārābī und Dominicus Gundissalinus', in *Albertus Magnus und der Ursprung der Universitätsidee. Die Begegnung der Wissenschaftskulturen im 13. Jahrhundert und die Entdeckung des Konzepts der Bildung durch Wissenschaft*, ed. by Ludger Honnefelder (Berlin: Berlin University Press, 2011), pp. 77–96
———, 'Religious Diversity and the Philosophical Translations of 12th-Century Toledo', in *Communities of Learning: Networks and the Shaping of Intellectual Identity in Europe, 1100–1500*, ed. by Constant J. Mews and John N. Crossley (Turnhout: Brepols, 2011), pp. 19–36
———, 'Dominicus Gundissalinus and the Introduction of Metaphysics into the Latin West', *The Review of Metaphysics*, 66 (2013), 691–712

———, 'The Arabic Influence on the Classification of Philosophy in the Latin West: The Case of the Introductions to Philosophy', *Micrologus: Nature, Sciences and Medieval Societies*, 28 (2020) (= *The Diffusion of the Islamic Sciences in the Western World*), 191–209

Gorochov, Nathalie, *Naissance de l'Université. Les écoles de Paris d'Innocent III à Thomas d'Aquin (v. 1200–v. 1245)* (Paris: Honoré Champion, 2012)

Grabmann, Martin, 'Die logischen Schriften des Nikolaus von Paris und ihre Stellung in der aristotelischen Bewegung des XIII. Jahrhunderts', in Martin Grabmann, *Mittelalterliches Geistesleben*, vol. 1 (Munich: Max Hueber, 1926), pp. 222–48

———, 'Magister Petrus von Hibernia. Der Jugendlehrer des heil. Thomas von Aquin. Seine Disputation vor König Manfred und seine Aristoteleskommentare', in Martin Grabmann, *Mittelalterliches Geistesleben*, vol. 1 (Munich: Max Hueber, 1926), pp. 249–65

Grant, Edward, *A Source Book in Medieval Science* (Cambridge, MA: Harvard University Press, 1974)

Grignaschi, Mario, 'Le *De divisione philosophiae* de Dominicus Gundissalinus et les *Quaestiones II–V in Sextum Metaphysicorum* de Jean de Jandun', in *Knowledge and the Sciences in Medieval Philosophy : Proceedings of the Eighth International Congress of Medieval Philosophy*, ed. by Simo Knuuttila, Reijo Työrinoja, and Sten Ebbesen, vol. 2 (Helsinki: Luther-Agricola Society, 1990), pp. 53–61

Hasse, Dag Nikolaus, 'Influence of Arabic and Islamic Philosophy on the Latin West', in *The Stanford Encyclopedia of Philosophy (Fall 2014 Edition)*, ed. by Edward N. Zalta <https://plato.stanford.edu/archives/fall2014/entries/arabic-islamic-influence/> [accessed 12 May 2020]

Höffe, Otfried, *Aristoteles* , 2nd edn (Munich: Beck, 1999)

Imbach, Ruedi, 'Einführungen in die Philosophie aus dem XIII. Jahrhundert. Marginalien, Materialien und Hinweise im Zusammenhang mit einer Studie von Claude Lafleur', *Freiburger Jahrbuch für Philosophie und Theologie*, 38 (1991), 471–93

Köhler, Theodor W., *Grundlagen des philosophisch-anthropologischen Diskurses im dreizehnten Jahrhundert. Die Erkenntnisbemühungen um den Menschen im zeitgenössischen Verständnis* (Leiden: Brill, 2000)

Lanza, Lidia, 'Artisten', in *Grundriss der Geschichte der Philosophie. 4/1–2: Die Philosophie des Mittelalters. 13. Jahrhundert*, ed. by Alexander Brungs, Vilem Mudroch, and Peter Schulthess (Basel: Schwabe, 2017), pp. 995–1001

Nederman, Cary J., 'Aristotelianism and the Origins of "Political Science" in the 12th Century', *Journal of the History of Ideas*, 52 (1991), 179–94

Rijk, Lambertus M. de, 'On the Genuine Text of Peter of Spain's *Summulae logicales*, III: Two Redactions of a Commentary upon the *Summulae* by Robertus Anglicus', *Vivarium*, 7 (1969), 8–61

Robiglio, Andrea A., '"Neapolitan Gold": A Note on William of Tocco and Peter of Ireland', *Bulletin de philosophie médiévale*, 44 (2002), 107–11

KATJA KRAUSE AND HENRYK ANZULEWICZ

Albert the Great's *Interpretatio*

Converting Libraries into a Scientific System

Introduction

Throughout the Middle Ages, Latin intellectuals invariably discussed the transmission and dissemination of knowledge by drawing on the two terms *translatio* and *interpretatio*. Unlike our contemporary view, which draws a clear distinction between translations and interpretations,[1] medieval ideals and practices of *interpretatio* were shaped by perceptions of continuity and

1 See, for instance, the summaries by Brentjes, Fidora, and Tischler, 'Towards a New Approach'; Wisnovsky and others, 'Introduction'. In their collected volume, Wisnovsky and others begin to reflect on the relationship and meanings of the terms 'transmission', 'translation', and 'transformation' (Wisnovsky and others, 'Introduction'), realizing that 'the categories of transmission, translation, and transformation are not entities in the natural world but, rather, enabling distinctions that allow us to assume different vantage points, and thereby to see the whole in greater depth and in sharper perspective' (p. 8). For an interesting collection on the topic of how translation products and processes were regarded throughout history, see Lefevere, ed., *Translation, History, Culture*. Important studies of the topic for the medieval period include: Brentjes, 'Reflexionen'; Burnett, 'The Institutional Context'; Burnett, *Arabic into Latin in the Middle Ages*; D'Alverny, 'Les traductions à deux interprètes'; Brentjes, Fidora, and Tischler, 'Towards a New Approach'; Freudenthal, 'Arabic and Latin Cultures'; Goyens, Leemans, and Smets, eds, *Science Translated*; Köpf and Bauer, eds, *Kulturkontakte und Rezeptionsvorgänge*; Speer and Wegener, eds, *Wissen über Grenzen*; Wisnovsky and others, 'Introduction'; Zonta, *Hebrew Scholasticism*. For a study that takes up contemporary approaches in engaging with Jerome's highly influential theory of translation and transliteration from Hebrew, see Hasselhoff, 'Hieronymus und die Aussprache'.

Katja Krause • (kkrause@mpiwg-berlin.mpg.de) leads the research group 'Experience in the Premodern Sciences of Soul and Body ca. 800–1650' at the Max Planck Institute for the History of Science, Berlin and professor (W2) for the history of science at the Technical University Berlin. Her research focuses on Latin medieval and early modern history of science and philosophy.

Henryk Anzulewicz • (anzulewicz@albertus-magnus-institut.de) is a faculty member of the Albertus Magnus Institute in Bonn, Germany, and editor of Albertus Magnus's writings. His research focuses on the history of ideas in the Latin Middle Ages, and in particular on Albertus Magnus.

Premodern Translation: Comparative Approaches to Cross-Cultural Transformations, ed. by Sonja Brentjes and Alexander Fidora, Contact and Transmission, 2 (Turnhout: Brepols, 2021), pp. 89–132

10.1484/M.CAT-EB.5.122135

association with translation. Rufinus of Aquileia's (345–410) Latin translations of Basil of Caesarea's (*c.* 329–379) eight sermons, for instance, were identified as *interpretationes* and not *translationes*, as we might have expected;[2] Burgundio of Pisa's Latin translation of John of Damascus's (670–756) *De fide orthodoxa* is referred to as *interpretatio Burgundionis* in a modern catalogue;[3] and Burgundio, in his translation of John Chrysostom's homilies on the Gospel of St John, referred to Jewish Septuagint translators as *interpretes*.[4] Burgundio's case, in particular, reveals that medieval intellectuals thought about the terms *translatio* and *interpretatio* in ways strikingly different from our own. Someone who translated literally (*ad verbum de verbo*), practising Boethius' omnipresent ideal, was, in Burgundio's eyes, a faithful translator (*fidus interpres*) and thus stood in opposition to a non-literal translator (*ad sensus de sensu*) who took 'the sense (*sententia*) of a matter' in order to 'explain it by the combination of [his] own words'.[5] More often than not, as Charles Burnett has eloquently shown, this non-literal method of translation was considered morally audacious, as it frequently provoked questions of authorship.[6]

These representative examples illustrate how the medieval concept of *interpretatio* extended from literal translation to the much wider processes of conveying sense and meaning. This becomes even clearer when we include reflections by medieval *interpretationes* that explore the full potential of the term. The *Glossa ordinaria bibliae*, for instance, collected glosses by the Church Fathers that were used *inter alia* as an aid in Biblical exegesis, were, as we would probably expect, referred to as *interpretationes* of the Bible. Just as this example includes a wider range of works, the personal term *interpres* too extended to commentators and autonomous authors alike, such as to a number of the Church Fathers and, in high Scholasticism, to the medieval sage Albert the Great (*c.* 1200–1280).[7] Key to our consideration here is that

---

2 See *Rufini Aquileiensis Homiliarum Basilii Magni interpretatio Latina*, ed. by Lo Cicero.

3 See John Damascene, 'Interpretatio latina Iohannis Burgundionis iudicis', BAV, MS Pal. lat. 314, fol. 187v–236v. Listed in the BAV as John Damascene, 'De fide ortodoxa, libri 1–4. Interpretatio latina Iohannis Burgundionis iudicis Pisani'.

4 See Burnett, 'Translating from Arabic into Latin'.

5 Burgundio of Pisa, *Commentary on John Chrysostom on the Gospel of St John Prologue*, Latin in Burnett, 'Translating from Arabic into Latin in the Middle Ages', p. 75; English: p. 57, amended: 'Si enim alienam materiam tuam tuique iuris vis esse putari, non verbo verbum, ut ait Oracius, *curabis reddere ut fidus interpres*, immo eius materiei sentenciam sumens tui eam dictaminis compagine explicabis'.

6 The boundaries between literal and non-literal translations also depended on the compatibility between the original and target languages. Whereas Greek was generally considered to lend itself to literal translations, the Arabic language was regarded as far more prolix and thus lent itself to explanation and clarity (*explanatio et claritas*). On this, see Burnett, 'Translating from Arabic into Latin', pp. 60–61.

7 See, for instance, Vosté, 'S. Albertus Magnus Evangeliorum Interpres'; Vosté, 'Sanctus Albertus Magnus Prophetarum Interpres'. Related medieval terms, usually featuring a narrower meaning included in *interpretatio*, are *translatio* (and cognates), *commentator* (and cognates), and *exegeta* (and cognates). For remarks concerning the terms *commentator* and

the general medieval Latin use of the term of *interpretatio* and cognates spanned a wide range of epistemic activities, including translations, exegeses, and explanations.

Whereas we commonly distinguish between these epistemic activities by means of different analytic terms in our histories of the transmission and dissemination of knowledge, medieval actors saw them as much more closely connected. Our aim in this essay is to complement the existing analytic histories by granting a voice to the historical actors themselves, and by historicizing the terms we use to analyse historical processes of knowledge transmission and dissemination.[8] Such actor-centred reflections by definition turn to the conceptual reality of the historical period, to the intentions and motives held by the historical actors in transmitting and disseminating texts, and to the

---

exegeta, see Günsz, *Die Abhandlung Alexanders von Aphrodisias*, p. 8; Moraux, *Alexandre d'Aphrodise*; Anzulewicz, 'Vermögenspsychologische Grundlagen', pp. 95–96; Falcon, 'Commentators on Aristotle'. The medieval term *translatio* is used in a similarly wide fashion as *interpretatio*, yet in some Scholastic thinkers it takes on a technical meaning similar to our contemporary use. This emerges very clearly from, for instance, Burgundio of Pisa's dedicatory epistle to his translation of Nemesius Emesenus' *De natura hominis*, p. 1: 'Dominatori Frederico, invictissimo Romanorum Imperatori et Caesari semper Augusto, a Burgundione, iudice natione Pisano, translatus incipit liber, anno incarnationis Domini MCLXV, indictione XIII. [...] Quia in meis, serenissime Imperator, Vobiscum locutionibus naturas rerum cognoscere et earum causas scire Vestram Maiestatem velle perpendi, idcirco librum hunc Sancti Gregorii episcopi Nyssae, fratris Sancti Basilii, de Graeco in Latinum Vestro nomine statui transferre sermonem [...]'. Similarly telling is Albert the Great's note in his commentary on Aristotle's *Politics*, II. 7. Ed. Paris. VIII, p. 163; see also Albertus Magnus, *Politica*, Berlin, Staatsbibliothek Preußischer Kulturbesitz, MS lat. 879, fol. 21rb, in which he refers to the translation of the pseudo-Aristotelian *Problemata*: 'Cuius ratio est in quodam libro *De problematibus quibusdam*, quem transtulit quidam David imperatori Friderico de Graeco in Latinum, et incipit: "Cum essem in Graecia, venit ad manus meas liber De problematibus Meteoron"'.

8 Previous interventions in the literature have embarked on connected paths, yet they have, to the best of our knowledge, analytically separated translations from other processes of knowledge transmission and dissemination. De Leemans, 'Aristotle Transmitted', for instance, has advocated a close relationship between a phase of textual translation and a phase of textual transmission, as he calls them. He suggests that both external and internal textual criteria grant us critical insights into the successes. Yet he limits these successes to a 'correct understanding of the translation' in the transmission phase, and a preservation 'of the original message' entailed in the text (pp. 333–35, 336, 338). For the remainder of the article, de Leemans proceeds to examine different remediation strategies employed by the readers of translations to make up for corrupt or unintelligible translations, on the assumption that the medieval readers' primary interest was to preserve 'the transmission of the original message' (p. 352). Similarly, despite separation, Wisnovsky and others have advocated for 'a synthetic cultural history' and emphasized the interrelatedness between the three processes of transmission, translation, and transformation, as they call them (Wisnovsky and others, 'Introduction'). In analysing the Latin and English etymologies of the term 'transmission', they note that the Latin root of *transfero* coincides with the English concept of transmission in a 'broader sense of textual motion and change' (p. 9). At the same time, they find an understanding of medieval translations to be impossible without reference to the terms 'transmission' and 'transformation' (p. 13).

meanings these actors ascribed to the transmitted and disseminated knowledge that they built into their own worlds. As we will discover, decisions concerning which corpora or libraries of knowledge to study, when to study them, how to study them, and to what end to study them were made by all actors involved in the activities of *interpretatio*. Hence, we consider the actors who explicitly identified as, or were identified as, *interpretes* — in all its meanings — to be as much involved in the medieval enterprise of transmitting and disseminating knowledge as those whom we commonly call translators.[9]

Our focus in this essay is on a, if not the, pivotal actor of interpreting (*interpretari*) the *corpus Aristotelicum* in Latin Scholastic thought, Albert the Great. Although he did not know Greek or Arabic, and was thus unable to engage in certain activities of *interpretatio* that modern actors would define as translation, we will seek to understand how Albert conceived of his own activity of *interpretatio* with regard to the diverse bodies of texts that became available in the Latin language shortly before and during his lifetime. Albert's acts of explaining, expositing, and commenting on the different scientific corpora and the libraries available to him, his endeavours to expand their meanings to a comprehensive scientific system and to inscribe into them a human teleology, all fell, in his eyes, under the range of the epistemic activities of *interpretatio* in which he was engaged. Indeed, just like his predecessors, Albert himself articulated the continuity between the activity of transferring a text from one language to another and the related explanatory activity by using the single term *interpretatio*, thus portraying a wide range of activities which uncovered, bridged, and forged meanings of and between terms, arguments, and even complete scientific corpora.[10] Hence we take Albert's understanding of *interpretatio* as the paradigm for our case study.

Our purpose is not, however, to trace the theoretical meaning that Albert assigned to the term, but rather to reconstruct his performance of *interpretatio* in and by means of the scientific system that he created, and to consider what this performance of *interpretatio* ultimately entails for

---

9 Here, we wish to expand upon the point made by Wisnovsky and others that it is the translator's doing alone 'to [...] *reconstitute* texts by transplanting them into the soil of a new culture, just as a scion is grafted to a growing plant whose rootstock binds it to a different terrain' (original emphasis; Wisnovsky and others, 'Introduction', p. 4).

10 The notion *interpretatio* (and its cognates) is pervasive in all of Albert's philosophical and theological works. Some exemplary passages where its twofold meaning is best revealed include: Albertus Magnus, *Meteora*, I. 4. 10. Ed. Colon. VI. 1. p. 41, 55–57; *Meteora*, III. 4. 3. p. 177, 1–4; *Meteora*, III. 4. 26. p. 202, 37–39; *Meteora*, IV. 1. 13. p. 226, 68–70; *Metaphysica*, V. 6. 1. Ed. Colon. XVI. 1. p. 282, 28–30; *De causis et processu universitatis a prima causa*, II. 1. 25. Ed. Colon. XVII. 2. p. 90, 90–94; *Super Matthaeum*, Ed. Colon. XXI. p. 508, 55–63; *De natura boni*, tr. 1. pars 2. c. 3 § 2, Ed. Colon. XXV. 1. p. 6, 57–59; *De natura boni*, tr. 2. pars 3. c. 1 § 2, p. 48, 44–49; *Quaestio de primis motibus*, a. 1. Ed. Colon. XXV. 2. p. 3, 12–17; *De bono*, V. 1. 4. Ed. Colon. XXVIII. p. 278, 4–10; *Summa theologiae*, I. 10. 43. 1. Ed. Colon. XXXIV. 1. p. 332, 2–6; *Super Dionysium De caelesti hierarchia*, c. 3. Ed. Colon. XXXVI. 1. p. 49, 34–39; *Super Dionysium De divinis nominibus*, c. 5. Ed. Colon. XXXVII. 1. p. 304, 82–85.

our histories of knowledge transmission and diffusion. Our principal focus is on two interrelated aspects that appear to be most striking in Albert's *interpretatio*: (1) the ways in which he forged a scientific system and the ultimate teleological purpose of human happiness that he connected to it; and (2) the path of his own previous scholarly activities, which led him to create these interrelated new meanings on the basis of his source texts. This entails addressing the question of complementarity between Albert's own scholarly expertise and the textual knowledge he encountered and transformed.[11] It also means that we understand how Albert used the corpora and libraries of his sources, the letters (*littera*), as he called them, from which he gathered the literal sense (*sensus litteralis*) in the cases of the sciences, and out of which he eventually derived the sentences (*sententia*) he composed.[12] Accordingly, we try to show how, and with which implicit intentions and motivations, Albert achieved a successful *interpretatio* of the knowledge contained in the *littera* in front of him through gathering their *sensus* and deriving their *sententia*. We will discover the extent to which he became a proponent of a new scientific system that was originally largely foreign to the world of Latin Scholasticism, but that, in and through his doing, would become a part of his own culture.

Forging a Scientific System

With his concluding remarks to the longest of his works, *De animalibus*, Albert put the final flourish to probably the most important contribution to natural philosophy (*scientia naturalis*) in and well beyond his time:

> Iam expletus est liber Animalium et in ipso expletum est totum opus naturarum in quo sic moderamen tenui quod dicta Perypatheticorum prout melius potui, exposui: nec aliquis in eo potest deprehendere quid ego ipse sentiam in philosophia naturali: sed quicumque dubitat, comparet hiis quae in nostris libris dicta sunt dictis Perypatheticorum, et tunc vel reprehendat vel consentiat me dicens scientiae ipsorum fuisse interpretem et expositorem: si autem non legens et comparans reprehenderit, tunc constat ex odio eum reprehendere vel ex ignorantia: et ego talium hominum parum curo reprehensiones.
>
> (The book of animals has now been completed, and with it, the entire work concerning natural things has been completed. In this, I have held as my governing principle to set forth the teachings of the Peripatetics to the best of my ability. It is not possible for

11 See Grabmann, 'Der Einfluss Alberts des Grossen', esp. pp. 345–52: 'Doctor universalis' and 'Doctor expertus'. See also Layer, 'Namen und Ehrennamen Alberts des Großen'.

12 See Robertson, *Lectio Divina*, p. 222.

> anybody to discern in this what I myself have perceived in natural philosophy. Yet, whoever has his doubts, may compare the things that have been said in our books to those of the Peripatetics. Then, he may either find fault or agree with me, and say that I was the one who interpreted and exposed their knowledge. If, however, he were to find fault without reading and comparing, then it is certain that he would find fault out of hatred or ignorance. Truly little do I care for the criticisms of such men!)[13]

No Scholastic thinker before or after him would comment on the entire body of Aristotle's natural philosophy, fill the gaps in the Stagirite's scientific programme with a large corpus of Peripatetic sources as well as with his own, autonomous works, and even integrate other non-philosophical disciplines such as theoretical medicine (*medicina theorica*) and pharmacology.[14] No other Scholastic thinker would erect an educational system that covered all the disciplines of rational (*scientia rationalis*), practical (*scientia practica*), and real philosophy (*scientia realis*), and whose effects would be discernible for centuries to come in the philosophical textbook traditions at the emerging universities, in medical curricula all over Europe, and in the erudition of young elites.[15] What Albert bequeathed to generations of his readers was nothing less than a complete body of philosophical knowledge — a body that he began to write in 1251 with his commentary on the most general discipline of natural philosophy, the *Physics*, and concluded after 1264 with his commentary on Aristotle's *Politics*. But what do Albert's epistemic reflections upon the creation of his scientific system tell us — reflections that surely could not have foreseen these long-lasting consequences?

One could suppose that Albert regarded his systematic exposition of all of Aristotelian natural philosophy as a creation of a bookish knowledge culture that focused on reiterating the teachings at the heart of his available sources. Yet nothing could be further from the truth. What Albert intended to pursue with his scientific system of natural philosophy, and equally so

13 Albertus Magnus, *De animalibus*, XXVI. § 36, ed. by Stadler, p. 1598, 7–15. Here and throughout, all translations are our own unless otherwise attributed. For the meaning of *scientia* in Albert (as we also use it throughout our contribution), and how it overlaps in meaning with *philosophia* for him, see, for instance, Albertus Magnus, *Physica*, I. 1. 1. – I. 1. 4. Ed. Colon. IV. 1. p. 1, 9 – p. 8, 13.

14 Concerning Albert's overarching system of natural philosophy in its structure, see Appendix I to this essay. Concerning his inclusion of medical knowledge into his science of the animals, see de Asúa, 'Minerals, Plants, and Animals'; de Asúa, 'War and Peace'; Siraisi, 'The Medical Learning of Albertus Magnus'; Schipperges, 'Das medizinische Denken'; Schipperges, 'Eine "summa medicinae"'; Krause, 'Grenzen der Philosophie'.

15 See, for instance, Honnefelder, ed., *Albertus Magnus und der Ursprung der Universitätsidee*, and particularly the introduction as well as the essays by Honnefelder, '"Bildung durch Wissenschaft"' and 'Metaphysik als "erste Wissenschaft"'; also Burger, 'Albertus Magnus'; Möhle, 'Albertus Magnus'; Donati, 'Alberts des Großen Konzept'; Anzulewicz, 'Alberts Konzept der Bildung'. See also Honnefelder, *Albertus Magnus und die kulturelle Wende*.

with his overarching system of philosophy, remains as concealed in the passage above as the Peripatetic writings did to the Latin audience before Albert's commentaries. For inasmuch as he was 'the one who has interpreted (*interpretem*) and exposed (*expositorem*) the knowledge (*scientiae*)' contained in his Peripatetic sources, there could, by virtue of the fact that it *is* knowledge (*scientia*), be no falsehood contained in his interpretation and exposition. Albert, therefore, was concerned not with interpreting and expositing teachings of opinion (*opinio*), but with transmitting perfect natural knowledge (*scientia naturalis perfecta*), as he had already clarified at the beginning of his *Physics*:

> Intentio nostra in scientia naturali est satisfacere pro nostra possibilitate fratribus ordinis nostri nos rogantibus ex pluribus iam praecedentibus annis, ut talem librum de physicis eis componeremus, in quo et scientiam naturalem perfectam haberent et ex quo libros Aristotelis competenter intelligere possent. Ad quod opus licet non sufficientes nos reputemus, tamen precibus fratrum deesse non valentes, postquam multotiens abnuimus, tandem annuimus et suscepimus devicti precibus aliquorum ad laudem primo dei omnipotentis, qui fons est sapientiae et naturae sator et institutor et rector, et ad utilitatem fratrum et per consequens omnium in eo legentium et desiderantium adipisci scientiam naturalem.
>
> > (It is our intention in natural philosophy to please — to the best of our ability — the brothers of our order. For many years now, they have been pleading that we may compose for them such a book of physical matters in which they would find perfect natural knowledge and whence they could understand the books of Aristotle competently. Although we do not consider ourselves sufficiently prepared for this work, we are nonetheless powerless to deny our brothers' supplications which we have declined many times already. At long last, we assent and, overcome by the supplications of some, we assume [this work], mainly for the glory of the omnipotent God, who is the fountain of wisdom and the author, creator, and ruler of nature, [and] also for the benefit of our brothers, and consequently of all who read this book and desire to acquire natural knowledge.)[16]

The language employed in this opening passage of Albert's *Physics* commentary reveals his systematic *interpretatio* of all of natural philosophy to be deeply inspired by his Aristotelian and Peripatetic commitments. The principal sake for which he composed his commentary was, as he states, 'the omnipotent God', whom he pertinently describes as 'the fountain of wisdom' and 'the author, creator, and ruler of nature'. God is thus portrayed as the single source of all knowing and of all being, and, at the same time, as the objective *telos*

16 Albertus Magnus, *Physica*, I. 1. 1. Ed. Colon. IV. 1. p. 1, 9–22.

of all knowing and being—a recurrent Neoplatonic motif which finds its expression in all of Albert's writings.[17] The secondary sake for which Albert penned his natural scientific works emerges in the idea of the benefit (*utilitas*) they will bring to his brothers and to all those who desire 'to acquire natural knowledge': in reliance on Aristotle's opening lines of the *Metaphysics*, Albert unmistakably regarded this desire as being inscribed in the natures of all human beings.[18] Yet here he suggests in addition that this desire is ultimately realized through the benefit (*utilitas*) that acquiring knowledge of the natural world can bring to his brothers. Indeed, the beneficial nature of natural philosophy, and by extension of all sciences, already hints at the worldview that Albert embedded into his scientific system, a system whose realization in the life of his brothers would have to follow the exact structure that Albert built into it.

This structure of Albert's scientific system was inspired by Aristotle's division of knowledge (*scientiae*), to be found in his *Physics*, II. 7. (198a 29–31) and *Metaphysics*, VI. 1. (1026a 18–19), and supplemented by additional works wherever Albert saw gaps in the Aristotelian system of knowledge.[19] Grounding Albert's structure, however, were commitments to Avicenna's and other Neoplatonic views on the human cognitive faculties. Indeed, for Albert what determined the ultimate foundation of the sciences was not the scientific objects, as would be expected in a purely Aristotelian system, but rather the human faculties of the senses and particularly of the intellect.[20] This different grounding shows the irrevocable link that he created between his scientific system and human nature, as he conceived of it. Real philosophy (*scientia realis*), be it concerned with natural, mathematical, or metaphysical objects, served as a path or instrument for developing and perfecting the human cognitive faculties and, in particular, the human intellect. At least in theory, it enabled humans to reach their ultimate goal of knowing the whole of reality in a divine-like way. Philosophy as a whole, Aristotelian knowledge (*scientia*), thus offered the tools for gathering the forms of nature, which Albert and his medieval peers perceived as particularized in the world, into the human mind. Any unwillingness to embark on this path or any ignorance that Albert certainly found in some of his brothers, as expressed at the end of the *De animalibus* passage cited above, amounted to their renunciation of the benefit (*utilitas*) that this knowledge granted in the development of their true nature as human beings. In Albert's eyes, therefore, this refusal could

---

17 See, for instance, Anzulewicz and Rigo, 'Reductio ad esse divinum'.

18 Aristotle, *Metaphysica*, I. 1. 980a21. See also Albertus Magnus, *Metaphysica*, I. 1. 4. Ed. Colon. XVI. 1. p. 6, 18–24 and 80–82; I. 24. p. 21, 87–p. 22, v. 7; II. 2. 2. p. 93, 81–p. 94, v. 6; XI. 2. 35. p. 527, 46–59.

19 See Jordan, 'Albert the Great', and Appendix I.

20 For primary literature, see Avicenna, *Philosophia prima*, tr. 1 c. 1, ed. by Van Riet, p. 1, 7 – p. 2, 30; *Logica*, cap. de intrando apud scientias, Venice 1508, repr. Frankfurt am Main 1961, fol. 2^{r}. For secondary literature, see Bertolacci, 'La divisione della filosofia'.

result in nothing but an ontological brutishness, an animal-like being, or an unrealized nature.[21]

For those, however, who did desire to know, and wished to benefit from Albert's system of the sciences in realizing their true nature as human beings, a particular order of study and an overview of the system were indispensable. Like most of his Scholastic contemporaries, Albert gave logic the first place in the order of study for young men, as it would train their thinking in the right way and help them avoid errors and falsities — pitfalls that stood at the beginning of the path to happiness.[22] This is the reason that Albert considered the study of logic beneficial (*utilis*) too, but this time in a different sense, namely in that it aided everyone to be equipped for the path of scientific inquiry:

> Est autem non tantum necessaria, sed etiam utilis haec scientia. Si enim bonum et felicitas hominis est secundum optimae partis animae hominis perfectissimum actum, hoc est secundum intellectum contemplativum, nec contemplari poterit intellectus, nisi noverit contemplationis principia et sciat invenire quod quaerit contemplari, et diiudicare id ipsum quod iam contemplatur inventum, patet quod prae omnibus utilis est ad felicitatem haec scientia, sine qua non attingitur felicitatis actus et per quam ipse felix actum non impeditae recipit operationis. Haec enim scientia a phantasiis, quae videntur et non sunt, liberat, errores damnat, et ostendit falsitates et lumen dat rectae in omnibus contemplationis. Prae omnibus igitur desideranda est haec scientia.
>
> > (For if the good of humans and their happiness lies in the most perfect activity of the best part of the human soul, [an activity] which lies in the contemplative intellect, this intellect would be unable to contemplate unless it had become acquainted with the principles and it would know to find what it searches for in contemplating, and [unless it knew] to judge the things that it has already found [when] it contemplates them. It is evident that this science — without which the activity of happiness remains unrealized and by which the happy man does not engage in the activity of unhindered happiness — is, above all [other sciences], beneficial for happiness. For this science liberates from fantasies, which seem to be but are not, it condemns errors and exposes falsities, and it supplies the light for a proper contemplation in all [its doings].)[23]

---

21 Albert repeatedly returns to this motif throughout his career. See, for instance, Albertus Magnus, *Super Dionysii Epistulas*, ep. 7. Ed. Colon. XXXVII. 2. p. 504, 27–32. At the end of every scientific corpus, Albert justifies the ways in which he used his libraries and rejects the ways in which his brothers received his works. See Albertus Magnus, *Metaphysica*, XIII. 2. 4. Ed. Colon. XVI. 2. p. 599, 61–66; *Politica*, l. 8 c. 6. Ed. Paris. VIII. pp. 803–04.

22 See note 15 above.

23 Albertus Magnus, *Super Porphyrium De V universalibus*, tr. de antecedentibus ad logicam, c. 3. Ed. Colon. I. 1A. p. 6, 16–29.

When Albert assembled his logical corpus, the availability of diverse libraries of logic — the old logical works (*logica vetus*) together with the new logical works (*logica nova*), multiple Latin and Arabic commentaries on them, and Avicenna's and Boethius' works — posed a unique challenge of *interpretatio*. How was it possible to synchronize these diverse works and to arrange them into a coherent system of *philosophia rationalis*, a system which would prepare the student for the study of the real sciences?[24] The *logica vetus* in its own right, which was part of the old *trivium* consisting of grammar, logic, and rhetoric, was a well-established corpus of texts. But when the logical works diversified, their synchronization and arrangement became a matter of resourceful interpretation. This is particularly visible in Albert's first commentary on rational philosophy, the one on Porphyry's *Isagoge*, which he prefaced with seven chapters, the so-called 'antecedents to the science' (*antecedentibus ad scientiam*). In these, he explained the nature of logic, its place within the overall philosophical corpus, its usefulness and necessity, its subject matter, and its parts.[25]

Yet Albert did not finish his synthesis of the logical libraries at this general level. On the contrary, in explicit reliance on Arabic insights propounded by Avicenna, Alfarabi, and Algazel, he divided his logical corpus 'on the basis of its intention', a division which for him entailed 'the teaching of principles by which one can reach, through things known, knowledge of the unknown'.[26] The unknown could either be simple, in which case logic had to treat the ways to reach definitional knowledge (*definitio vel notificatio*), or complex, in which case it had to treat the ways to attain demonstrative knowledge through syllogistic reasoning (*argumentatio*). In this respect, the twofold division of Albert's *philosophia rationalis* was nothing out of the ordinary, and yet it presented his distinctive response to the Latin problem of synthesis. Just as he grounded Aristotelian knowledge (*scientia*) as a whole in the human cognitive faculties, so Albert built his unified logical corpus on the different intellectual activities and thus skilfully intertwined the library of the *logica vetus*, to which he ascribed the intellectual activities of ordering (*ordinare*), composing

24 See Tremblay, 'Albert le Grand'.

25 Albertus Magnus, *Super Porphyrium De V universalibus*, tr. de antecedentibus ad logicam, Ed. Colon. I. 1A. pp. 1–2: Chapter 1, 'An logica sit scientia specialis'; pp. 2–4: Chapter 2, 'An logica sit aliqua pars philosophiae'; pp. 5–6: Chapter 3, 'Ad quid necessaria est logicae scientia et ad quid utilis'; pp. 6–8: Chapter 4, 'De quo sit logica ut de subiecto'; pp. 8–11: Chapter 5, 'De divisione logicae et partibus eius et de significationibus dictionum'; pp. 11–14: Chapter 6, 'De parte logicae, qua certificatur incomplexum, et partibus eius et modis'; pp. 14–16: Chapter 7, 'De parte logicae qua certificatur questio de complexo'.

26 Albertus Magnus, *Super Porphyrium De V universalibus*, tr. de antecedentibus ad logicam, c. 5. Ed. Colon. I. 1A. p. 8, 28–32: 'Divisio autem logicae et quae sunt partes ipsius, ut dicunt Avicenna et Algazel et Alfarabius, accipienda sunt ex intentione ipsius. Sicut vero iam ante dictum est, logica intendit docere principia, per quae per id quod notum est, deveniri potest in scientiam ignoti'. For an excellent discussion in the secondary literature, see Brumberg-Chaumont, 'Les divisions de la logique'.

(*componere*), collecting (*colligere*), and dividing (*dividere*),[27] with that of the *logica nova*, to which he linked the single activity of resolving (*resolvere*).[28]

Following this systematic division by way of intellectual activities, Albert thus integrated all the logical libraries at his disposal. The *logica vetus* which he commented upon as a whole included Porphyry's *Isagoge*, Aristotle's *Categories* (in Boethius' translation), Gilbert of Porretanus' *De sex principiis*, Boethius' *De divisione*, and Aristotle's *De interpretatione* (again in Boethius' translation). The *logica nova* comprised Aristotle's *Analytica priora* (in the translations by Boethius and anonymous), Aristotle's *Analytica posteriora* (in James of Venice's translation),[29] the *Topica* (in Boethius' translation),[30] and the *De sophisticis elencis* (in the translations by Boethius and James of Venice).[31] The libraries at his disposal thus took on a systematic meaning that had not been granted to them before in any previous Scholastic *interpretatio*.

Equally important was Albert's integration of numerous insights of Arabic provenance into his commentaries on the whole body of *philosophia rationalis*. For instance, he prefaced his commentary on Aristotle's *Analytica posteriora* with a paraphrase of Algazel's *Logica*, similar to his prefacing the *Liber de causis* with Algazel's *Metaphysics*, and he drew on Arabic sources wherever the template texts fell short.[32] At the end of his *philosophia rationalis*, Albert reflected on the extent to which his labours for completeness and order inaugurated a unique *interpretatio* which he endowed with an explicit legacy.

> Sic enim de rhetoricis (ut dictum est de rhetoricis) erant quidem multa et antiqua dicta et bona: de syllogismis autem per artem traditam nihil habuimus prius ante nos. Unde nec de tota logica. [...] nisi exempla quaedam particularia, ex quibus in attritione exercitii et adaptationis quaerentes similia ex similibus, multum tempus laboraremus sine arte quaerentes ea quae sunt artis, quod per artis compendium cito potest fieri. Si autem videtur considerantibus ea quae hic tradita sunt, quod sufficienter haec ars habeat ea omnia quae a principio erant proposita dicere de arte syllogistica, et habere super alia negotia in Gorgia et Zenone

27 See Albertus Magnus, *Liber divisionum*, tr. 1 c. 1. Ed. Colon. I. 2. p. 81, 5–15.

28 See Albertus Magnus, *Super Porphyrium De V universalibus*, tr. de antecedentibus ad logicam, c. 7. Ed. Colon. I. 1A. p. 15, 15–41, and c. 5. p. 8, 28–53.

29 Aristotle's *Analytica posteriora* was also available in the translations made by Ioannes and Gerardus of Cremona, but these were not Albert's templates.

30 Aristotle's *Topica* were also available in a translation made by anonymous, but, again, this translation was not Albert's template.

31 Concerning the translators and translations of the logical works from Greek and Arabic into Latin, see Dodd, 'Aristoteles Latinus', esp. pp. 74–75; Trizio, 'Greek Aristotelian Works Translated into Latin'.

32 See Albertus Magnus, *Analytica posteriora*, I. 1. 2. Ed. Paris. II. pp. 4–7; see also 'Logica Algazelis Introduction and Critical Texts', cap. de materia syllogismi, ed. by Lohr, pp. 273–80; Albertus Magnus, *De causis et processu universitatis a prima causa*, I. 1. 10. Ed. Colon. XVII. 2. pp. 18–24, *Algazel's Metaphysics*, ed. by Muckle, p. 1 tr. 2, pp. 52–61.

exemplariter tradita, et supra ea negotia quae ex dictione Antiquorum et posteris adaucta sunt, reliquum ex dictis relictum erit omnium vestrum qui haec quae tradita sunt legeritis, et eorum qui audierint hoc opus exponi ab aliis, de omissis quidem artis a nobis qui adjutorium nullum habuimus, tradere indultionem sive indulgentiam, inventis autem sive de inventis multas habere sive agere nobis grates.

(Of rhetorical matters, there have indeed been many old and good pointers. Yet of syllogisms, transmitted by way of an art, we had nothing before our times, and thus neither of logic as a whole [...] except for some detached examples, out of which, scrutinizing examples from examples in the sweat of my labour and revision, I worked long hours scrutinizing the things which are of an art without an art — [a procedure] which, by way of a compendium of this art, can be carried out swiftly. Yet if it seems to those who ponder the things that have been treated here that this art sufficiently holds everything it proposed at the beginning to say of a syllogistic art and to preserve other matters treated by Gorgias and Zeno in exemplary fashion, and matters which were augmented by the pointers of the ancients and their followers, there will remain a residue for you all from the pointers that have endured. [This will remain for] all of you who will read what has been transmitted, and for those who will listen to this work [when it is] explained by others. Of those things, however, that have been omitted by us of this art — we who had no aid — it is [for you] to grant clemency and indulgence. Yet, in the things that have been found and of the things that have been found, it is [for you] to honour much and to carry on in gratitude to us.)[33]

Once a student had acquired the complete body of knowledge available in the *philosophia rationalis* that Albert thus bequeathed to his successors, he was sufficiently prepared to embark on the study of all other Aristotelian knowledge (*scientia*). Yet each of the following sciences, according to Albert, had its own way of knowing, 'in accordance with the diversity of the subject matter in which knowing is sought'.[34] Logic, however, served as method for acquiring all the real and practical sciences (*modus omnis philosophiae*) on whose basis human happiness would eventually be attainable. This comment in particular reveals how Albert considered the relation of logic to the goal of human happiness. Logic, in the sense of an autonomous science, as *philosophia*

33 Albertus Magnus, *De sophisticis elenchis*, II. 5. 3. Ed. Paris. II. pp. 712b–13b. For Albert's knowledge of Zeno's contribution to logic, see Albertus Magnus, *De praedicamentis*, tr. 1 c. 2. Ed. Colon. I. 1B. p. 6, 54 with note 54, apparatus of sources.

34 Albertus Magnus, *Super Porphyrium De V universalibus*, tr. de antecedentibus ad logicam, c. 7. Ed. Colon. I. 1A. p. 15, 45–47: 'Hic tamen modus (sc. accipiendi scientiam de quolibet scibili incomplexo vel complexo) secundum materiam, in qua ponitur, variatur secundum diversitatem materiae, in qua quaeritur scientia'.

rationalis per se, was instrumental for acquiring the sciences that lead to happiness: it was a preparatory knowledge to be acquired. In its applied sense within the other sciences, in its beneficial (*utilis*) or performative meaning, logic steered and directed the human intellect to become all things through scientific inquiry, and to ultimately attain to happiness.[35] Logic, therefore, assumed a dual character in Albert's scientific system: it was preliminary to all sciences as an object of study, as *philosophia rationalis*, and it was applied in all of them as a tool of study, as a method that facilitated knowledge acquisition and ultimately human happiness.

Parallel in its duality yet somewhat different in meaning was the role that Albert assigned to ethics (*philosophia moralis*) in his overarching scientific system.[36] As an object of study, this practical science educated the student about virtue, the good life, and human happiness (*ethica docens*). It thus stood in close relation to the other practical sciences of economics and politics.[37] As a tool in its performative value for life, however, ethics equipped every philosopher with indispensable traits of moral virtue (*ethica utens*). In its theoretical value, ethics thus revealed the content of human happiness, for the civil life or the contemplative life, but in its practical value, it stipulated moral virtue as a necessary requirement for the contemplative life.[38] Just like logic, therefore, the performative value of ethics did not lead to ultimate happiness in its own right, but was rather a requirement for it. Yet unlike logic, indeed unlike any other Aristotelian science, ethics occupied a unique space of reflexivity in Albert's scientific system. It was the science that was able to desegregate all isolated scientific pursuits, pursuits that were separated by their subject matters and their particular methods, and integrate them into the greater whole of the benefit (*utilitas*) that they carried for the human being as human being.[39] The rational science of logic and the practical science of ethics, therefore, received two unique *interpretationes*. On the one hand, as practices pursued in the service of scientific inquiry, they both facilitated the contemplative path that could eventually lead to happiness; on the other hand, as sciences pursued in their own right, they were subjects indispensable for acquiring complete knowledge of the whole of reality.

---

35 Albertus Magnus, *Super Porphyrium De V universalibus*, tr. de antecedentibus ad logicam, c. 3. Ed. Colon. I. 1A. esp. p. 5, 63–73.

36 See Albertus Magnus, *Super I Sententiarum*, d. 3 a. 16. Ed. Colon. XXIX. 1. p. 89, 46: scientia naturalis, rationalis et moralis; *Summa theologiae sive de mirabili scientia dei*, I. 3. 13. 3. Ed. Colon. XXXIV. 1. p. 43, 30–34: physica, logica, ethica. In this passage, logic and ethics are explicitly considered in parallel fashion.

37 See, for instance, Albertus Magnus, *Super Ethica*, prolog., Ed. Colon. XIV. 1. p. 2, 51–58; *Super Ethica*, prolog., Ed. Colon. XIV. 1. p. 2, 67 – p. 3, 16; *Ethica*, I. 1. 6. Ed. Paris. VII. p. 16a; *Metaphysica*, I. 1. 1. Ed. Colon. XVI. 1. p. 3, 15–17.

38 See, for instance, Albertus Magnus, *Super Ethica*, prolog., Ed. Colon. XIV. 1. p. 3, 18–24; *Metaphysica*, XI. 1. 9. Ed. Colon. XVI. 2. p. 473, 10–24.

39 See Avicenna, *Logica*, cap. de intrando apud scientias, Venetiis 1508, fol. 2^{r}. Albertus Magnus, *De homine*, Ed. Colon. XXVII. 2. p. 4, 49–58.

In contrast to the dual characters of logic and ethics, Albert took the real sciences (*scientiae reales*), under which he subsumed natural philosophy (*philosophia naturalis*), mathematics (*mathematica*), and metaphysics (*philosophia prima*), as directly conducive to man's ultimate happiness. In this beneficial sense, these sciences transcended their enclosed truths and aspired to a greater meaning, which Albert expressed in the perfecting effect that they had on the human intellect. Physics, mathematics, and metaphysics all facilitated the slow and laborious conjoining of the two intellectual parts possessed by each human with each other, and thus gave rise to the coming-to-be of an intellectual nature in the full sense of the word, which resulted in intuitive knowing of all intelligible truths already in this life.[40] Yet while Albert circumscribed happiness as the beneficial aspect of the real sciences, he considered the proximate goals of every science to be necessary. Indeed, this distinction between benefit and necessity preserved the autonomy of the sciences as sovereign bodies of knowing with their own principles, subject matter, and intrinsic teleologies.[41]

Scientific autonomy did not, however, exclude connection and alliance. Albert's natural philosophy, especially his *De animalibus*, went hand in hand with his *Metaphysics* in a programmatic way. The methodological principle of his natural philosophy was that 'we have married metaphysics with natural things, so that this teaching be more perfect and the things said more easily understood'.[42] Albert repeatedly communicated this marriage in the axiomatic formula 'the work of nature is the work of the intelligence' (*opus naturae est opus intelligentiae*),[43] as much as in his understanding of nature as one of the

---

40 See, for instance, Albertus Magnus, *De anima*, III. 3. 11. Ed. Colon. VII. 1. p. 222, 15–26.80–84, and c. 12, p. 221, 90 – p. 225, 5.

41 Concerning the beneficial nature (*utilitas*) of the *scientiae*, see, for instance, Albertus Magnus, *Super Porphyrium De V universalibus*, tr. de antecedentibus ad logicam, c. 3. Ed. Colon. I. 1A. p. 6, 16–29 as quoted in note 23; *Super Ethica*, VI. lect. 10. Ed. Colon. XIV. 1. p. 463, 48–78. Müller eloquently discusses the relationship between philosophy and theology and the autonomy of each of the philosophical sciences on the basis of their proper scientific principles, subject matter, methods, and purposes (Müller, *Natürliche Moral und philosophische Ethik*, pp. 30–37).

42 Albertus Magnus, *Liber de natura et origine animae*, tr. 2. c. 17. Ed. Colon. XII. p. 44, 15–23: 'De his tamen omnibus primi philosophi determinare est opus. Sed in eis quae hic diximus, cum naturalibus metaphysica composuimus, ut perfectior sit doctrina et facilius intelligantur ea quae dicta sunt; haec enim est consuetudo nostra in toto hoc physico negotio. De generatione igitur et distinctione et natura et statu animae et incorruptione et opere tantum dicum est a nobis'.

43 See, for instance, Albertus Magnus, *De animalibus*, XVI. 1. 7. ed. by Stadler, p. 1082, n. 43: 'Est autem hiis adhuc addendum quod cum omne opus naturae sit opus intelligentiae, sicut in aliis libris physicis iam determinavimus, quod istud maxime est opus intelligentiae in quo virtus intelligentiae opus naturae terminat ad formam sibi propinquiorem et similiorem. Cum enim unum et idem sit opus naturae et intelligentiae eo quod natura per intelligentiam et intelligentia in natura operatur: terminus tamen operis aliquando propinquior est naturae et aliquando propinquior est intelligentiae. Quando enim formae quae secundum esse et

four Neoplatonic primary causes (*causae primariae*) superseding the four Aristotelian causes:

> Quattuor autem sunt causae primariae in una ratione causalitatis sumptae. [...] Tales enim non secundum se primariae possunt esse. Ratio enim earum non repugnat, quin materiales sint. Materiales autem primariae esse non possunt. Quattuor autem sunt: causa prima, intelligentia, anima nobilis et natura, prout natura forma diffusa est in primis naturantibus principiis omnia constituentibus in incohatione et motu et perfecto esse naturali.
>
> > (The primary causes, however, are four, congregated in one nature of causation. [...] For such [causes] cannot be primary according to themselves. But [the primary causes] are four: the first cause, the intelligence, the noble soul, and nature — nature, in this sense, is the diffusive form in the first nature-granting principles in all constituents at the beginning, in motion and in perfection of natural being.)[44]

The primary cause of nature in the Neoplatonic sense of the principle of being, motion, and perfection which Albert promulgates in this passage is all-diffusive and pours its formal abundance in an all-pervasive fashion over the whole of reality. Nature in the narrower sense as perfection entails, once again, that the benefit of Albert's scientific system is grounded in a shared ontology between the cosmos as a whole and the human being with its cognitive and noetic capacities. What is diffused form throughout the cosmos, extending too to the human soul and intellect, can be mentally restored to its concentrated form, to form per se, within the human intellect and thus inform the intellect in its return to its true intellectual nature. Only through a comprehensive inclusion of the forms of reality into the intellect is this ontology reinstantiated naturally.[45] This is the ultimate reason that, in Albert's eyes, all sciences have to be studied in the proper order and comprehensively mastered in order

---

posse sunt materiales et corporeae sunt fines huiusmodi operis, pro certo est terminus operis propinquior naturae et vincunt in opere qualitates elementales primae et consequentes quae causantur ex eis in mixto. Quando autem elevatur forma super conditiones materiae corporalis, secundum quod plus et plus elevatur, sic plus et plus appropinquat ad intellectualis naturae sive intelligentiae similitudinem, et vincunt tunc vires intelligentiae in naturae opere'.

44 Albertus Magnus, *De causis et processu univ. a prima causa*, II. 1. 2. Ed. Colon. XVII. 2. p. 62, 2–14.

45 See, for instance, Albertus Magnus, *Metaphysica*, I. 5. 15. Ed. Colon. XVI. 1. p. 88, 60 – p. 89, 84, esp. p. 89, 55–60; 11 tr. 1 c. 9. Ed. Colon. XVI. 2. p. 472, 87 – p. 473, 55, esp. p. 473, 10–15: 'Cum igitur lumen intellectus et intellectuum, sicut iam dictum est, sparsum sit in omnibus intelligibilibus, primum, quod acquirit intellectus per studium, est collectio intellectuum speculativorum in tribus theoricae partibus, quae sunt physica, mathesis et prima philosophia'. See also Albertus Magnus, *De anima*, I. 1. 1. Ed. Colon. VII. 1. p. 2, 52–p. 3, 5; III. 3. 11. p. 221, 81 – p. 222, 9; *De intellectu et intelligibili*, II. c. 12. Ed. Paris. IX. p. 520a–21b; *Liber de natura et origine animae*, tr. 2 c. 13. Ed. Colon. XII. p. 38, 42 – p. 39, 12.24–46.

to lead the intellect back to its nature and source. Albert's passage from the beginning of his *Physics*, the twofold purpose and worldview it communicates, can now be read in light of these considerations: God as the creator and *telos* of all reality, particularly of intellectual natures, has inscribed into them the capacity to return to their true natures and thus to Him by way of studying the comprehensive body of the sciences.

In Albert's insistence on the criterion of comprehensiveness, too, natural philosophy played a particularly prominent role, forming the largest part of Albert's scientific system. Beginning with the most general of the natural sciences, *Physics*, passing through the body of knowledge concerned with the inanimate nature in its expression in simple and complex bodies, and eventually turning to the sciences covering the animate nature in its expression in souls and bodies, Albert expanded the library of Aristotle's natural philosophy by altogether eighteen books. In full awareness of the gaps that the Stagirite had left especially in this part of *scientia realis* and that the Greek and Arabic Peripatetic and medical sources had revealed to him, Albert reconstructed the missing knowledge either from these sources or autonomously. His natural science of place (*De natura loci*), for instance, was written with only partial reliance on Julius Honorius' *Cosmographia*; his natural science of nutrition (*De nutrimento et nutrito*) was a self-standing work that attempted to reconstruct Aristotle's lost work with the same title; and his natural science of the intellect (*De intellectu et intelligibili*) featured the knowledge propounded by illustrious thinkers such as Alexander of Aphrodisias, Alfarabi, Avicenna, and Eustratius.[46]

Albert's requirement of comprehensiveness was not, however, exhausted in possessing a complete system of all natural and, by extension, real sciences that could be had. Especially in the natural sciences, Albert insisted on the specificity of knowledge as an additional fundamental criterion of comprehensiveness. All physical things, whether inanimate or animate, ought to be known in their maximal specificity, even if this knowledge concerns vile and disgusting things.[47] Details and factual knowledge about natural things (*historia*), therefore, mattered as much for Albert's completion of the natural scientific corpus as did its breadth and scope. As he expounded at the beginning of his *Physics* commentary, it was both aspects that led to perfect natural knowledge.

Once again, Albert linked these two aspects of the criterion of comprehensiveness to the development and noetic ascent of the human intellect: all natural sciences followed the order of discovery, from knowing indistinct simple things to knowing distinct composite things, an intellectual movement which, in Albert's eyes, was perfectly mapped onto the mental activity of

---

46 See Appendix II.

47 See, for instance, Albertus Magnus, *De animalibus*, XI. c. 1. ed. by Stadler, pp. 761–64, esp. p. 761, 13–29.

composition (*compositio*). As Albert explains in his *Physics*, this movement initiates in the vague abstractions of sensory experience and ends in knowing definite particulars of all physical beings.[48] It is in this mental operation of composition, in particular as performed in the natural sciences, that we see Albert's explanations from his *philosophia rationalis* take on their practical value, benefiting the human intellect in its noetic ascent through assimilation to all forms of reality.

In performing the two epistemic activities of interpreting and explaining, as we witnessed them at the outset of this section, Albert forged a comprehensive scientific system, consisting of the rational, practical, and real science. They carried new meanings in their ideals and practices that transcended Aristotle's original intentions as well as those of his manifold sources. Albert's commitment to perfect knowledge in the natural sciences (*scientia perfecta naturalis*) entailed an intrinsic scientific value and a beneficial worth, both of which by far surpassed a simple encyclopaedic ordering of the sciences and a plain obligation to acquire demonstrative knowledge as the intrinsic *telos* of the Aristotelian sciences. Any student who followed the order of his scientific system would acquire true knowledge about the given part of reality he studied in a particular science (an ideal which Albert regarded as the intrinsic value of each science), and potentially would, in addition, reach true human happiness by perfecting his intellect through studying these sciences (an ideal which comprised the beneficial value of all sciences taken together in the effects they bore on the human intellect).

What is more, these values of an active pursuit of the sciences in the real lives of those who study them expose the ways in which Albert engaged with and drew upon his sources. Notwithstanding their Aristotelian, Greek and Arabic Peripatetic, or Latin provenances, he integrated the knowledge (*scientia*) contained in them in a realist and not in a bookish way. In keeping with the epistemic standards for any *scientia* worthy of the name — truth and certainty, to name the two most important — Albert took every acquired item of knowledge to amount to a real impression of the world's formal reality onto the mind. His harmonization of distinct bodies of knowledge of one and the same science (as exemplified in his *philosophia rationalis*), his vision of logic and ethics as sciences that carried theoretical and practical values at once, his integration of extensive paraphrases of Arabic sources, and his criterion of comprehensiveness (as we saw them conveyed in Albert's *philosophia naturalis*) all express this realism in distinct yet essential ways. For these structures were all carefully tailored to yield intellectual knowledge through scientific discovery, and, in their effects, to assimilate the human intellect to the truths contained in the world. The methods that Albert suggested for scientific discovery ranged from the experiential to

48 See, for instance, Albertus Magnus, *Physica*, I. 1. 6. Ed. Colon. IV. p. 10, 5–18.

the doctrinal to the rational, and they were always in accordance with the science that was studied.

In this and many other ways, Albert thus created a scientific system that was intended to last for centuries in its systematic order and scientific value. In its very ontology, it was grounded in his vision of a teleological anthropology, meant to bring into intellectual being the *interpretatio* which Albert had inscribed into his scientific system. While many generations of Latin Scholastic intellectuals adopted some of the *interpretationes* that Albert put forth in his scientific system, most of them strikingly rejected its overall teleological force for their own particular reasons.[49] Nonetheless, Albert's overall *interpretatio* of a new and complete scientific system achieved an unprecedented and unrepeated inculturation of Aristotelian and Greek and Arabic Peripatetic knowledge during a Latin epoch that had slowly begun to realize the significance of this foreign knowledge for its own identity, but was still hesitant to accept the worldview it carried.

Forging the Scientific *Telos*

Albert's *interpretatio* as put forth in his mature scientific system, together with its overarching motives, accomplishments, and aims, speak the language of successful integration of Peripatetic knowledge into this system on a large scale. The existential performance of his scientific system, and the insights about human nature and its *telos* upon which that system rested, shaped Albert's choices of what subjects to study, how to study them, and which ideals to apply in these scientific pursuits. These *interpretationes* and the new meanings of human knowing and teleology they introduced, however, have their own history, which, in Albert's own thinking and doing, predated his mature vision of a scientific system by at least two decades.

As will be shown in this section, the very possibilities that enabled Albert to erect his system upon a performative anthropology and teleology find their initial expressions in his earliest theological works. While these seemingly have little to do with his final vision of a systematic philosophy, they offer us a glimpse of his interconnected and gradual arrival at the existential insights with which he endowed the final version of his scientific system. Originally, Albert was committed to a traditional image of Latin theology, focusing on morality. But he soon shifted the focus of his systematic thinking on human life's teleological dimension from a moral pursuit to an intellectual one. How exactly he arrived at the view that such an intellectual pursuit, as we have seen it inscribed into his mature scientific system, could lead humans to their fulfilment is the theme of this section.

---

49 See, for instance, Steel, *Der Adler und die Nachteule*, esp. pp. 26–29, 42–43; Jordan, 'Albert the Great', pp. 484–85; Anzulewicz, 'Albertus Magnus und seine Schüler', pp. 191–92.

From the first stroke of his pen to the last, Albert anchored his thought in one stable anthropological *telos*, the idea of perfection. His first treatise, *De natura boni*, promulgated this perfection in the theological vision of human imitation of divine substantial goodness. What applied to God *simpliciter* pertained to humans derivatively. Following the lead of Philip the Chancellor's (Philippus Cancellarius) *Summa de bono*, Albert's early anthropology set out to attend to human character and behaviour in light of Church doctrine and eschatology, and portrayed human perfection in terms of the good of human nature, human action, and the different divine graces given to humans. Indeed, it set out 'to treat the differences of [human good] in a moral way' and 'to describe the nature of this created good in this book more morally than substantially'.[50] The view of human teleology Albert propounded in this work featured moral actions in their circumstances as perfection. Albert's understanding of the teleology of humans in the first two parts of his *De natura boni* thus took up a limited perspective on social aspects alone: good intentions and right actions, civil virtues and chastity were the paths that led to human perfection in this life.[51] Yet the part on human nature was never realized in this work, and, as a consequence, Albert did not consider the question of what is natural about humans. Nonetheless, he did conceive of an order of nature along traditional motives found in Augustinian and Boethian theological triads. In explaining how these triads, innate to every creature, pave the way to the Good per se, Albert conceived of human nature as literally 'ordained and led by its hand to God' (*manuductio*).[52]

Soon thereafter, Albert set aside his attempts to explain human teleology through traditional moral theology alone, although he remained deeply committed to an overarching theological enterprise. His *De IV coaequaevis*, a work dedicated to discussing the six days of creation on the basis of its four coequal principles of matter (*materia*), time (*tempus*), the heavens (*caelum*), and the angels (*angelica natura*) and written around 1242/1243, bears witness to a new era in Albert's view of human nature and perfection. Against the backdrop of a new theological framework, that of the revived pseudo-Dionysian theology, Albert began to read human nature and teleology in its place in the ontological hierarchy, at the intersection of the transcendent and contingent realms of being, and in their psychological

50 Although Albert's initial plan of this work was to cover the whole of creation and recreation, he was unable to realize it in his *De natura boni*. For this original plan, see Albertus Magnus, *De natura boni*, Ed. Colon. XXV. 1. p. 1, 5–9, 16–23. Instead, Albert concentrated on interpreting the good of circumstances (*bonum circumstantiae*) as part of the good of political virtue, and even here it is remarkable that he increasingly deviated to considering chastity (*castitas*) on the basis of Mariology. Ultimately Albert gave up the path of writing his *De natura boni*.

51 See Albertus Magnus, *De natura boni*, tr. 2 pars 3 c. 1 § 2. Ed. Colon. XXV. 1. p. 36, 10–13 and 37–42.

52 See Albertus Magnus, *De natura boni*, tr. 1 pars 1. Ed. Colon. XXV. 1. p. 1, 34–41 and 49–59.

disposition to return to God as their principle.[53] Importantly, these new commitments coincided with the rise of Albert's insistence on inquiring into the most fundamental constituents of reality. Questions about the essence or whatness of all there is soon permeated all of his learned endeavours, just as they did those of his contemporaries in their ceaseless attempts to integrate the new knowledge system of Aristotelian philosophy.[54] Albert's inquiry into the kinds of resemblance between angels and humans in his *De IV coaequaevis* perfectly set the scene for promulgating this new approach to human nature and teleology, insofar as it integrated pseudo-Dionysian theology with Aristotelian psychophysiology and Augustinian psychology. The central meaning that Albert bestowed upon the varied source texts in this way was that of their mutual compatibility, merging seemingly disparate understandings of a theme into an integrative system that was capable of weighing their implications accordingly. But how exactly did Albert proceed to create such an integrative system in his *De IV coaequaevis*?

Humans, he reasoned there, resemble angels in intellectual virtue, rational knowing, and free will, but less so in physical appearance, and least of all in their sense powers, since these belong to brute animals.[55] This systematic insight fused in equal measure pseudo-Dionysian ideals, on what is naturally constitutive of human nature and what sets humans apart from brute animals, with Aristotelian and Avicennian paradigms that 'the ultimate and principal form in nature [...] is the rational soul'. With Augustine, Albert subsequently divided this conception of soul into a threefold structure of simple intellect, compositive intellect, and free will. Finally, he mapped the working of the simple intellect onto the complete corpus of Aristotelian *scientiae*, and suggested that 'every power of the intellect is affirmed as a power for rational, ethical, natural, mathematical, and divine things'.[56] This passage, in particular,

---

53 See Albertus Magnus, *De IV coaequaevis*, tr. 4 q. 34, Ed. Paris. XXXIV. p. 522; a. 2, p. 524b; q. 52, p. 589a. It is worth mentioning that the treatise on the angels in Albert's *De IV coaequaevis* is a question commentary on pseudo-Dionysius' *De caelesti hierarchia*; hence it is of little surprise that pseudo-Dionysius is one of Albert's main sources. For secondary literature, see Anzulewicz, *Die theologische Relevanz des Bildbegriffs*, pp. 42–48.

54 See Rodolfi, '"Ex nihilo id est post nihilum"'; Twetten 'Albert the Great's Early Conflations', p. 62; Rodolfi, 'La dottrina della materia celeste', esp. pp. 29–34; Rodolfi, *Il concetto di materia*; Winkler, 'Seele–Engel–Intelligenzen'; Roling, *Locutio angelica*, pp. 73–78; Rosier-Catach, 'The Augustinian Threefold Word', esp. pp. 26–30; Chung, 'Dissociating Angels and Intelligences'.

55 See Albertus Magnus, *De IV coaequaevis*, tr. 4 q. 61 a. 4. Ed. Paris. XXXIV. pp. 655b–56a.

56 Albertus Magnus, *De IV coaequaevis*, tr. 4 q. 61 a. 4. Ed. Paris. XXXIV. pp. 655b–56a: 'Quoddam enim inest homini gratia corporis: ut sursum habere intuitivas virtutes, et figurae rectum et luculentum [...] Quoddam autem inest homini gratia animae, ut intellectuale. Quoddam autem gratia conjuncti inest, sicut secundum naturam principale et regale. Et dicitur principale secundum naturam, eo quod formam ultimam in natura et principalem, quae est anima rationalis, attingit. Et dicitur regale propter dominium et regimen suorum actuum. Id autem quod praestat irrationabilibus animalibus, est triplex. Quorum primum inest secundum intellectum simplicem: quia licet minimum sit quoad

foreshadowed the grounding of his scientific corpora in the equivalent faculties of the human soul. In his *De IV coaequaevis*, however, Albert did not yet think through the implications of his statement, and instead pressed on with his ontological comparison of humans to angels. Accordingly, he suggested that humans chiefly 'agree with angels' in their simple intellectual nature, and metaphorically in their rational intellectual nature, even though angels 'possess constitutive and rational knowledge not by investigation of seeking and searching, but according to the order of the things known'.[57] Angels differed in their inborn mode of knowing (in part because they were actual intellects, whereas humans were potential intellects), but not in their psychological ontology.

This tentative vision of a continuity in intellectual nature between angels and humans would become pivotal in Albert's full-fledged development of human teleology in his later works. The first step in his *De IV coaequaevis*, however, consisted in refocusing his considerations of human nature and teleology along cosmological and psychological lines. Albert no longer located the good of human nature in morals alone, that is, in good intentions and right actions, but rather in their intellectual soul. What is constitutive of every human being as human, Albert forcefully repeats at the very end of *De IV coaequaevis*, is their 'rational nature alone'.[58] Some years later, in commenting on the *corpus Dionysiacum*, this insight would find its reformulation in the famous principle, cited above, that humans inasmuch as they are humans are their intellect alone, which he used from then on.[59] Yet this reformulation was hardly different from its original. Angels, in Albert's *De IV coaequaevis*, were just said to be as 'rational' as humans, meaning of course that they shared in the same 'intellectual' nature, as just discussed. And intellectual humans were indeed, in his eyes — to the extent that all scientific knowing was an

---

sensibiles virtutes, in quibus convenit cum brutis: tamen omni virtute est potens secundum magnitudinem intellectus. Et dicitur omnis virtus intellectus virtus ipsius ad rationalia et ethica et naturalia et mathematicalia et divina. Secundum inest per intellectum compositivum, qui ratio est, secundum quod omni continuitate habet scientiam rationalem: continuat enim terminum termino in habitudine subjecti ad praedicatum, et habitudine propositionis ad propositionem secundum quamlibet speciem ratiocinationis. Tertium est in ratione liberi arbitrii, scilicet secundum naturam animae liberum et potentissimum'.

57 Albertus Magnus, *De IV coaequaevis*, tr. 4 q. 61 a. 4. Ed. Paris. XXXIV. p. 656a: 'et [angeli] habent continuitivas et rationales scientias non per investigationem inquisitionis, sed secundum ordinem rerum scitarum'.

58 Albertus Magnus, *De IV coaequaevis*, tr. 4 q. 69 a. 3 particula 3 quaesitum 1. Ed. Paris. XXXIV. p. 703a: 'homo enim in eo quod est homo, tantum est rationalis naturae: et sic sensualis motus est extra ipsum, id est, extra naturam hominis in eo quod homo'.

59 This formula in all its expressions throughout Albert's career is discussed in Anzulewicz, 'Anthropology: The Concept of Man in Albert the Great', pp. 340–46; Anzulewicz, 'Solus homo est nexus Dei et mundi', pp. 330–35; Köhler, *Grundlagen des philosophisch-anthropologischen Diskurses*, pp. 585–624.

explicit expression and reflection of their intellectual soul — an expression that animals in their nature did not share with humans.[60]

Albert's insight into scientific knowing as an expression of the human intellectual nature did not stop at integrating theoretical commitments in a theological framework, nor did it entail a comprehensive library at his disposal at this time, a library stretching from Augustine's *De diversis quaestionibus 83*, *De Genesi ad litteram*, and *De trinitate* to pseudo-Dionysius' corpus, to Boethius' *De consolatione philosophiae*, the *Liber de causis*, Alfred of Sareshel's *De motu cordis*, Isaac Israeli's *De definicionibus*, Avicenna's *Sextus De naturalibus* and *De caelo et mundo*, Moses Maimonides' *Liber de uno deo benedicto*, Averroes' *De caelo et mundo*, *De substantia orbis*, and *Metaphysics*, and John of Damascus's *De fide orthodoxa*. Rather, already at this early stage of Albert's thinking, it entailed, in the same way as the moral theological teachings of his *De natura boni* and their implementation in homilies and everyday monastic life, the manifestation of a lived reality. The repercussions that these theoretical insights into human nature had on real life, that is, on the lives of Albert and his envisioned audience, were their implementation and existential performance in and through scientific work. Without the ideas as transmitted in his libraries, Albert would have never been in a position to review his insights into human nature. Yet this review required as much his own ideals as the ideas contained in his libraries. Albert's vision, therefore — his first steps in penetrating into the very core of human nature as foretold in his *De IV coaequaevis* — was the first milestone on the path to forging a scientific system capable of fulfilling human nature in reality.

The second milestone was the comprehensive examination of human nature, and Albert undertook this in minute detail in *De homine*, the second part of his monumental *Summa de creaturis*. His enthusiasm for jumping head first into this investigation rested on an expressed concern to attain a comprehensive understanding of human nature in itself and in its psychophysiological totality.[61] This entailed nothing less than an examination of all faculties of the human soul, their corporeal organs and their functions, and, last but not least, their distinctive objects.[62] Albert recorded this study against the backdrop

60 Albert the Great's earliest division of the sciences and the human capacity for them is found in his *De IV coaequaevis*, tr. 4 q. 61 a. 4. Ed. Paris. XXXIV. p. 656a, as quoted above in note 56.

61 While he refreshed his comparison between human and angelic nature, now explicitly combining Aristotelian with Augustinian views, Albert conceived of the human soul as the act of the body in contrast to the angelic intellect, which he took to be a substantial form, and suggested at the same time that the human soul is made in the image of God in a special agreement of external things in addition to its internal agreement in the faculties of memory, intellect, and will, which it also shared with the angels. See, for instance, Albertus Magnus, *De homine*, Ed. Colon. XXVII. 2. p. 31, 40–44, and p. 546, 22–48.

62 Albertus Magnus, *De homine*, Ed. Colon. XXVII. 2. p. 1, 2–5: 'Consequenter transeundum est ad quaerendum de homine. De quo primo quaerendum est secundum statum eius in seipso et postea de loco eius, qui paradisus dicitur. Circa statum eius in seipso quaerenda sunt tria.

of all the sources at his disposal at the time, and in this way composed an anthropological study of a depth and breadth unprecedented in the Latin West, one that would only seldom rely on his Latin contemporaries and their limited appropriations of the new sources.[63] Albert now began to integrate into his comprehensive examinations Aristotle's psychological works, headed by *De anima* and the *Parva naturalia*, together with newly translated works of the Greek and Arabic Peripatetics, such as Alexander of Aphrodisias, Alkindi, Avicenna, Algazel, and Averroes. Augustine's, Boethius', and Gundissalinus' psychological works were of less importance, although still quite present.[64]

A major challenge that Albert faced in formulating this first fully fledged anthropology was the need to adapt the natural philosophical foundations of human action found in Aristotle and the Greek and Arabic Peripatetics in order to fit them in with the equally modified theological commitments he found in the Platonizing moral psychology of Augustine and other theologians. At the outset of his *De homine*, Albert thus reiterated the importance of pseudo-Dionysius, now in conjunction with Plato, and emphasized a philosophical understanding of the human soul as mover of the body.[65] The systematic breakthrough for conciliating these sources, however, came through the generic concept of 'psycho-motor powers' (*vires motivae*). In this concept, Albert joined theological views with philosophical ones, and ordered them according to hierarchical ideals. The image of God in the soul (*imago dei*) took the place of the highest power, as it conformed human actions to God's exemplar.[66] The Aristotelian conceptions of three psycho-motor powers, the practical intellect (*intellectus practicus*), the imagination (*phantasia*), and the appetitive power (*appetitus*) complemented his reading of this *imago dei*. Albert thus proposed a basic understanding of human nature and teleology

---

Quorum primum est de anima ipsius, secundum de corpore, tertium de coniuncto'.

63 Albert's new approach was based on reflections upon previous approaches to the Aristotelian and Peripatetic sources by Philip the Chancellor, Alexander of Hales, Robert Grosseteste, and William of Auvergne. Albert himself reflects on his renunciations in the following passages: Albertus Magnus, *De homine*, Ed. Colon. XXVII. 2. p. 401, 21, p. 412, 49–68. Reflections by Albert's predecessors on their approaches to the new sources are found, for instance, in Alexander Halensis, *Summa theologica*, II. Ed. Quaracchi. p. 446sq., esp. pp. 453–55; Philippus Cancellarius, *Summa de bono*, ed. by Wicki, p. 270, v. 197sq.; Guilielmus Alvernus, *Tractatus De anima*, Ed. Paris. II. 2. p. 205b sq., 210a sq. In the secondary literature, Albert's renunciations of certain teachings have already been discussed. For Grosseteste, see, for instance, Davidson, *Alfarabi, Avicenna, and Averroes*, p. 215, and n. 456; for William of Auvergne's theory of the intellect, see, for instance, Schneider, *Die Psychologie des Albertus Magnus*, pp. 185–86; Jüssen, 'Wilhelm von Auvergne', p. 163. Davidson, *Alfarabi, Avicenna, and Averroes*, p. 215.

64 See Index 'Auctores ab Alberto ipso allegati', in Albertus Magnus, *De homine*, Ed. Colon. XXVII. 2. pp. 18–26, esp. pp. 24–26, and Index 'Auctores a nobis allegati' (pp. 637, 639, 642–45).

65 Albertus Magnus, *De homine*, Ed. Colon. XXVII. 2. pp. 18–26, esp. pp. 24–26; see also Anzulewicz, 'Anthropology'; Anzulewicz and Krause, 'Albert der Große'.

66 Albertus Magnus, *De homine*, Ed. Colon. XXVII. 2. p. 504, 2–5; also pp. 539–55; *Quaestio de synderesi*, a. 1. Ed. Colon. XXV. 2. p. 233, 57 – p. 234, 13, esp. p. 234, 8–13.

that was meant to last. In his *De anima* he reiterated this view, but without using the theological language of *imago dei*.[67]

His reflections on the intellect of humans also substantiated the new generic conceptualizations of human nature. The threefold division through which he approached the human intellectual nature in *De homine* — the psychological configuration of the intellect and its developmental stages, the intelligible object of the intellect, and the relation of the intellectual to the sensitive soul — became the gold standard of his later works.[68] His *De intellectu et intelligibili*, the work in which he would take up the theme of human perfection in its final elaboration, rested on the very same division as his *De homine*.[69] Through the hundreds of folios on these themes, Albert merged and formalized the vocabulary, structures, and approaches spread across his source texts.[70] The threefold intellectual configuration that he built in reliance on his Arabic Peripatetic sources — the agent intellect, possible intellect, and speculative intellect — remained as the basis of subsequent elaborations on intellect and opened up possibilities for integrating further developmental stages, such as the actual intellect and the divine intellect. Especially in connection with his insights into the relation between intellect and intelligible object, in his *De intellectu et intelligibili*, Albert would settle definitively on each stage of human intellectual *perfectio*, a process for which he drew heavily on the newly available source of Alfarabi's *De intellectu et intellecto*. Already in his *De homine*, however, Albert revealed that he knew how to think about human nature against the backdrop of Aristotelian and Peripatetic libraries. Merging key ideas contained in them into a coherent whole and combining them with theological thought of different origins, he proceeded by following Aristotelian standards of methodology from the simple to the more complex matter and from the things more known to us to those more knowable in themselves. Yet he did not consider human teleology in light of his new-found Aristotelian commitments. Its philosophical formulation and integration would have to await the more mature periods of his thinking, when it would find its theoretical *interpretatio* in its most developed form in Albert's *De intellectu et intelligibili* and *De natura et origine animae*, and its

67 In his *De anima*, Albert considers these *vires motivae* again in a digression where he now portrays them as an extension of the philosophical viewpoint of Aristotle. However, Albert here does not make any reference to the *imago dei*, now interpreting the *vires motivae* no longer theologically but rather epistemologically. Albertus Magnus, *De anima*, III. 4. 10. Ed. Colon. VII. 1. pp. 240–42.

68 Albertus Magnus, *De homine*, Ed. Colon. XXVII. 2. p. 396, 2–3: 'Deinde quaerendum est de intellectu. Et quaeruntur tria, scilicet de differentia intellectuum, et de differentia intelligibilis, et de comparatione intellectus ad sensibilem animam'.

69 See Albertus Magnus, *De homine*, Ed. Colon. XXVII. 2. pp. 396–442: 'De intellectu'; 'De differentia intelligibilis', pp. 443–54.

70 See Hasse, 'Das Lehrstück von den Vier Intellekten', esp. pp. 62–67; Anzulewicz, 'Entwicklung und Stellung der Intellekttheorie', pp. 182–87; Sturlese, *Vernunft und Glück*, pp. 5–9.

application throughout his scientific system.[71] But in his *De IV coaequaevis* and *De homine*, Albert had already begun thinking about human nature in its fundamental relevance by means of philosophical insights, and thus laid down all the constituents required to develop a philosophical teleology of human nature.

The first place where Albert developed his philosophical views on human teleology, however, was the moral context, the same context in which his vision of a comprehensive system of human knowing and acting had originated. In the second treatise that Albert composed on moral goodness, the *De bono* (*c.* 1243 in Paris), he briefly parted ways with the study of the intellectual nature of humans and resumed an earlier interest: investigating the Good as the cause of all of creation and as participated in by everything created. Recalling his earliest work, the *De natura boni*, Albert insisted again that any created good can only be understood in relation to the uncreated Good, which is God. Contrary to that work, though, Albert now limited the moral lens in its application from the whole of reality to the four cardinal virtues and their onto-theological foundation in the Good. With this limitation, he reconceived of the nature of the Good and how humans participated in it. The increasingly influential body of Aristotelian works, in this case the *Nicomachean Ethics*, set new scientific ideals to pursue that reconceptualization, ideals which were hardly reconcilable with the original project built on Philip the Chancellor's *Summa de bono*.[72] The 'new ethics' Albert advocated in his *De bono* no longer conceived of the Good as a self-diffusive object that is desired by all creation. This unique combination of pseudo-Dionysius' *bonum diffusivum sui* with Aristotelian desire dynamized Albert's view of the Good: God who diffused himself in humans' nature was to be aspired to in nature's perfection, at least

---

71 The nature of humans as the basis for their teleological perfection emerges even in the structure of Albert's mature works, *De intellectu et intelligibili* and *Liber de natura et origine animae*. In the first book of his *De intellectu et intelligibili*, Albert initially treats of the nature of the intellect (book I. tr. 1: *De natura intellectus*) then of its object, the intelligible (book I. tr. 2: *De per se intelligibili*), and subsequently of the relation of the two to one another (book I. tr. 3: *De comparatione intellectus ad intelligibile*). In the second book of his *De intellectu et intelligibili*, Albert focuses on the intellectual development and natural intellectual perfection of the human being as human (book II. tr. unicus: *De perfectione naturali intellectus*). Albert structures his *Liber de natura et origine animae* in a similar fashion: its first treatise is concerned with the nature of the soul in the body, while the second treatise considers the perfected nature of the soul in the afterlife. See Albertus Magnus, *De intellectu et intelligibili*, Ed. Paris. IX. pp. 477–521; *Liber De natura et origine animae*, Ed. Colon. XII. pp. 1–44. In the secondary literature, this is discussed in Hornig, 'Perfektibilität'; Anzulewicz and Rigo, 'Reductio ad esse divinum'.

72 See Albertus Magnus, *De bono*, tr. V. q. IV (addita), Ed. Colon. XXVIII. pp. 300–07. This added question to his *De bono* is a strong indication of Albert's developing understanding of the ethical virtues in Aristotle's *Nicomachean Ethics*. This, of course, goes hand in hand with the developing availability of this work in the Latin West together with Averroes' middle commentary on it. For a discussion of these matters, see Müller, *Natürliche Moral und philosophische Ethik*, pp. 156–62.

to the extent that it was possible for this particular nature.[73] The ethical dimension in its limitation to ethics in Albert's *De bono*, therefore, proved to be one key ingredient in engineering a view of human nature as the first dimension to require natural perfection. A second key ingredient in *De bono* was that Albert began there to contemplate the relation between prudence and science, which catalysed his deliberations on the relation between human science and teleology.[74]

The breakthrough application of human teleology in the realm of the intellect and its development came, as mentioned above, in Albert's first full commentary on Aristotle's *Nicomachean Ethics*, his *Super Ethica*. In this programmatic writing, composed at the Dominican *studium generale* in Cologne between 1250 and 1251, Albert now revisited the topic of ultimate human happiness from a purely philosophical viewpoint.[75] Earlier, in his commentary on pseudo-Dionysius' *De divinis nominibus*, he had already set the tone in this direction by reconceiving contemplation as conducive of ultimate happiness.[76] In book X of his *Super Ethica*, then, it was the link that Albert forged between contemplation and the speculative intellect that established the psychological effect of ultimate happiness. Albert thus no longer portrayed human intellectual nature in its psychological ontology alone, no longer as a status quo, as he had still done to a great extent in his *De homine*, but now merged it with the developmental perspective he had originally devised in the ethical realm in *De bono*, and envisioned it explicitly as first and second perfection.[77] At the same time, he supported his new emphasis on a developmental psychology of human nature with a panoply of insights about ultimate happiness from his libraries of different traditions. Aristotle's *Nicomachean Ethics* (x), Boethius' *De consolatione philosophiae* (iii. met. 12), Averroes' long commentary on the *Metaphysics* (xi. comm. 51), the Byzantine commentators of Aristotle's *Nicomachean Ethics* (anonymous and Michael of Ephesus), pseudo-Dionysius' or rather Avicenna's *Philosophia prima* (viii. c. 7), and Anselm of Canterbury's *De casu diaboli* (c. 4) were all explicitly invoked. But their support was not simply sought to back up the new ideal. On the contrary, just as in his *De natura boni* and *De bono*, Albert found that these sources confirmed his *interpretatio* of human nature and teleology in

73 See Albertus Magnus, *De bono*, tr. 1 q. 1 a. 1. Ed. Colon. xxviii. p. 1, 11–12; also p. 4, 53–58.

74 See Albertus Magnus, *De bono*, Ed. Colon. xxviii. p. 226, 35–51.

75 See Albertus Magnus, *Super Dionysium De caelesti hierarchia*, c. 10. Ed. Colon. xxxvi. 1. p. 166, 29–42.

76 See Albertus Magnus, *Super Dionysium De divinis nominibus*, c. 4. Ed. Colon. xxxvii. 1. p. 133, 71–73. In his somewhat later *De anima*, Albert will introduce a further developmental stage of the human intellect, the so-called acquired intellect (*intellectus adeptus*), and will place human happiness in this developmental stage. See *De anima*, iii. 3. 11. Ed. Colon. vii. 1. p. 222, 21–28; c. 12 (p. 224, 90 – p. 225, 9). For secondary literature, see Anzulewicz, 'Entwicklung und Stellung der Intellekttheorie', pp. 188–93.

77 See Albertus Magnus, *De anima*, iii. 2. 7. Ed. Colon. vii. 1. p. 187, 48–55.

its practical and existential dimensions as well. The acquired intellect that Albert thought humans could reach naturally, or the *perfecta operatio intellectus speculativi,* as he called it in *Super Ethica,* was not simply a piece of theory, but presented the goal that was inscribed potentially in Albert's own motivating powers of affect and desire:

> Bonum enim maxime est illud quod per se bonum est et ad quod alia ordinantur, et hoc praecipue est perfecta operatio intellectus speculativi, quae dicitur felicitas contemplativa.
>
> (For that which is good is maximally that which is the Good in itself. And it is to this [Good] that all things are ordered, and that which is first and foremost [ordered to it] is the perfect activity of the theoretical intellect, which is said to be contemplative happiness.)[78]

Albert's ideal of ultimate human perfection was thus immediately put into practice in his own life. The path leading to this ultimate goal of self-perfective activity, which he had already discovered in his *De IV coaequaevis* and developed further in his later works, such as *Metaphysics,* was to pursue a system of the sciences that would cater to that perfection, not only in his own life, but also in the lives of all those who would follow him.[79] Indeed, it was the practice of commenting on the sciences and acquiring them, the practice of *interpretatio,* that in Albert's eyes constituted the key to reaching the human goal of *perfectio.* This he underscored in his commentary on Aristotle's *De anima,* writing that 'we study by discovering and by listening to teachers', and thus acquire theoretical knowledge through this activity. The causal effect of this knowledge, however, is to increasingly assimilate the possible intellect to the agent intellect, in such a way that it becomes 'more and more similar to itself day by day' until it has fully acquired its form. Albert did not see these Peripatetic insights as a theoretical piece of doctrine, but rather as an insight into the reality of human nature, an insight so true for him that it became one of the deepest commitments of his lived anthropology. In the final stage of intellectual development, once all sciences had been mastered fully, Albert took the perfected human being to perform 'his very own activity inasmuch as he is human, and this is the activity that is performed by God, which is to contemplate perfectly through one's own [powers] and understand separate substances'.[80]

78 Albertus Magnus, *Super Ethica,* x. lect. 1. Ed. Colon. xiv. 2. p. 708, 4–15. See also Albertus Magnus, *De natura boni,* tr. 1 pars 1, Ed. Colon. xxv. 1. p. 1, 4–6 and 28–31.

79 See Albertus Magnus, *Metaphysica,* v. 4. 1. Ed. Colon. xvi. 1. p. 272, 14–27. For a discussion of this matter in the secondary literature, see De Smet '"Perfectio prima–perfectio secunda"'. That essay notes Albert's importance to these two notions only in passing and with regard to his *Summa theologiae,* ii. a text of dubitable authenticity.

80 Albertus Magnus, *De anima,* iii. 3. 11. Ed. Colon. vii. 1. p. 221, 73 – p. 222, 28: 'Quaedam autem speculata fiunt in nobis per voluntatem, quia scilicet studemus inveniendo et audiendo a doctore, et haec omnia fiunt intellectu agente influente eis intelligibilitatem, et faciendo haec

Naturally, in this picture, moral development and *perfectio* are subsumed under the intellectual, not only because the intellectual realm is that which determines the specific human nature, but also because it is granted methodological significance for all scientific endeavours. His peers' reluctance to study the scientific system that he had given them with this perfective *telos* in mind — a reluctance we encountered at the beginning of this paper — was thus, in Albert's eyes, nothing less than a reluctance to realize their own *perfectio* as human beings.[81] The education set in place by the Dominicans and the universities served this teleological purpose, the full realization of a nature that humans already possess at the outset of their lives (*prima perfectio a natura*).[82] But Albert interpreted this process of realization as a development through various intellectual stages, stages that required the study of all the different sciences and resulted in different relations of humans to reality as a whole.

Conclusion

With Albert's scientific system, which, as we have seen, included commentaries on the *corpus Aristotelicum* along with his own independent works, the *doctor universalis* envisioned and contemplated the realization of a teleology inscribed in human beings in real life.[83] With this vision, he did

intellecta secundum actum esse intellecta intellectus agens coniungitur nobis ut efficiens. Et quia omnibus his influit intellectualitatem et denudationem, sunt omnia sibi similia in hoc quod separata sunt et nuda; et ideo in omnibus his accipit continue intellectus possibilis lumen agentis et efficitur sibi similior et similior de die in diem. Et hoc vocatur a philosophis moveri ad continuitatem et coniunctionem cum agente intellectu; et cum sic acceperit omnia intelligibilia, habet lumen agentis ut formam sibi adhaerentem, et cum ipse sit lumen suum, eo quod lumen suum est essentia sua et non est extra ipsum, tunc adhaeret intellectus agens possibili sicut forma materiae. Et hoc sic compositum vocatur a Peripateticis intellectus adeptus et divinus; et tunc homo perfectus est ad operandum opus illud quod est opus suum, inquantum est homo, et hoc est opus, quod operatur deus, et hoc est perfecte per seipsum contemplari et intelligere separata ut formam sibi adhaerentem, et cum ipse sit lumen suum, eo quod lumen suum est essentia sua et non est extra ipsum, tunc adhaeret intellectus agens possibili sicut forma materiae. […] Et tunc homo perfectus et divinus effectus est ad suum opus, inquantum homo et non animal est, perficiendum'. Parallel insights are expressed in Albertus Magnus, *De intellectu et intelligibili*, II. c. 12. Ed. Paris. IX. p. 520a–b.

81 See Albertus Magnus, *Metaphysica*, V. 4. 1. Ed. Colon. XVI. p. 272, 19–27; see also *De intellectu et intelligibili*, II. c. 10. Ed. Paris. IX. p. 518a–b. For a discussion of this matter in secondary literature, see Anzulewicz, 'Entwicklung und Stellung der Intellekttheorie', pp. 165–71.

82 See Albertus Magnus, *De anima*, II. 3. 3. Ed. Colon. VII. 1. p. 100, 15–22.

83 See Albertus Magnus, *Liber de natura et origine animae*, II. 13. Ed. Colon. XII. p. 39, 8–23: 'Propter quod prae omnibus beati philosophi studentes primo in physicis venerunt postea ad mathematica et ultimo se ad studium divinorum vertentes in eis vitam finiverunt. Et hanc felicitatem prae omnibus aliis Solon Atheniensis commendavit dicens animam post mortem in his felicitari. Propter quod cum in ultimo spiritu esset

not see himself as delivering a new system of knowledge that would replace Church teachings, render obsolete Biblical and systematic theology, or raise philosophy above competing systems of sciences. He also did not see himself as simply commenting upon a random library, spreading bookish learning of a different sort than before the entry of Aristotelian works in the Latin West, or making the new knowledge contained in these books available to his brothers for any use whatsoever. Rather, Albert regarded himself as an interpreter and expositor of Peripatetic knowledge (*scientiae*) defined by true and certain insights into reality, and thus as facilitating the realization in their intellectual nature of all humans who were willing to embark on the path of studying that knowledge.

Very much like the moral knowledge of the earlier theological tradition, Albert regarded Peripatetic knowledge as providing a compass, guidance, even a rule as to the path through which to attain the ultimate human goal. But unlike older moral knowledge, Peripatetic knowledge contained in its scientific approach the very existential route that would lead upwards towards the ultimate goal of returning to one's fount on the basis of one's nature. Everything propounded in this scientific system seemed, in Albert's eyes, conducive to obtaining that goal: practical knowledge (*prudentia*) in ethics, economics, and politics controlled the intra- and inter-human levels; understanding (*intelligere*) the *corpus Aristotelicum* afforded theoretical knowledge; and finally, acquisition of the whole body of Peripatetic *scientia* yielded the attainment of the ultimate goal inscribed in human nature. However, the universal path Albert had in mind for all humans to reach this goal, a path located in his scientific system, could only be realized individually, in the personal pursuit of acquiring the knowledge contained in his system every day of one's life. Albert's intention, therefore, was nothing less than to naturalize (as Abdelhamid Sabra put it) the whole body of Peripatetic knowledge, but to do so as its well-versed interpreter and expositor.[84] Significantly, his self-identification as interpreter of Peripatetic *scientia* and expositor of the truth contained in the libraries transmitting this knowledge joined his forging of a scientific system and shaping of a scientific *telos* with all other activities of *interpretatio*.

Albert's epistemic practice of *interpretari* encompassed a large range of epistemic objects: at times, it turned to notions and terms, theological opinions and philosophical arguments, Latin texts and translated writings from Greek and Arabic as its matter of inquiry. At other times, it focused on entire bodies

---

constitutus et Metrodorus et quidam alii coram eo conferrent de quodam intelligibili divino, supra vires collecto spiritu caput erexit et cum a Metrodoro compatienti sibi rogaretur caput ad cervical submittere, dixit: Sinite, ut et hoc intelligibile de quo confertis, apprehendam, antequam moriar, ut ampliori luce felicitatis elevetur intellectus. Hic igitur est status animae rationalis post mortem felicissimus in luminibus istis secundum naturam'.

84 Sabra, 'Appropriation and Subsequent Naturalization of Greek Science'.

of knowledge only to culminate in the creation of an entire scientific system. This was precisely the object of *interpretatio* that we have discussed in this essay. Nonetheless, it was not the epistemic objects investigated, studied, or examined that were the decisive factor of demarcation in Albert's practice of *interpretatio*, but the practice of *interpretari* itself, coupled with its epistemic motivation. Just as it had manifold objects of inquiry, *interpretari* was not a singular epistemic act in its applied expression, such as the act of explaining the meaning of a term or translating a text from one language to another. On the contrary, *interpretari* extended to a panoply of intricately related performances of discursive reasoning (*ratiocinatio*) and forging meaning which included, but were not limited to, translating, explaining, understanding, inferring, foretelling, determining, and concluding. What held these various discursive acts of *interpretari* together was Albert's singular epistemic motivation of eliciting truth and certainty through them.

Albert's *interpretatio*, in its conceptual and its performative meanings, thus illustrates a different way of seeing how epistemic objects, epistemic activities, and epistemic actors can be integrated wholes in processes of knowledge making, transmission, and dissemination. For Albert, these processes did not consist of separated activities and isolated practices alienated from each other by seemingly impermeable boundaries, as when we tend to refer to them separately by terms such as transmission, translation, or appropriation, to name just a few activities described in the literature. The idea of a fundamental continuity between these processes, which Albert and his peers expressed unmistakably through his conception and performance of *interpretatio*, can highlight unity while still seeing diversity. By applying Albert's *interpretatio* and employing it as a tool for our analysis of knowledge dissemination studies, we have begun to appreciate the particular medieval practices that he performed through the historical lens of an actor's category.

Appendix I: Albert's Division of his Natural Science (*scientia naturalis*) according to Books and Order as Discussed in his *Physica*, I. 1. 4

Book [Edition]	Subject Matter	Template / Main Source
1. ***Physica*** [Alberti Magni Opera omnia, Ed. Colon. IV. 1–2]	*corpus mobile in se absolute simpliciter et universaliter consideratum*	1. Arist., *De auditu physico*, trans. vetus 2. Ps.-Arist., *De lineis indivisibilibus*, trans. Rob. Grosseteste
2. ***De caelo et mundo*** [Alberti Magni Opera omnia, Ed. Colon. V. 1]	*corpus mobile ad situm sive circulariter sive in rectum consideratum in se*	Arist., *De caelo et mundo*, trans. vetus (Gerardi Cremonensis)

Book [Edition]	Subject Matter	Template / Main Source
3. ***De longitudine et latitudine terrarum et civitatum et de locis habitabilibus*** [*De natura loci*, Alberti Magni Opera omnia, Ed. Colon. v. 2]	*corpus mobile ad situm secundum quod habet ordinem ad mobile secundum formam, quod recipit impressiones eius in habitudine, quam habet ad locum generationis impressionum illarum*	Julius Honorius (Ps.-Aethicus), *Cosmographia* (only in tr. 3).
4. ***De causis proprietatum elementorum et planetarum*** [*De causis proprietatum elementorum*, Alberti Magni Opera omnia, Ed. Colon. v. 2]	*corpus mobile ad situm secundum quod habet ordinem ad mobile secundum formam, quod recipit impressiones eius*	Ps.-Arist., *De proprietatibus elementorum*, trans. Gerardi Cremonensis
5. ***De generatione et corruptione*** [Alberti Magni Opera omnia, Ed. Colon. v. 2]	*corpus mobile ad formam, simplex et contractum ad materiam*	Arist., *De generatione et corruptione*, trans. vetus (Anon.)
6. ***Libri Meteororum quattuor*** [*Meteora*, Alberti Magni Opera omnia, Ed. Colon. VI. 1]	*corpus mobile commixtum in via mixtionis*	Arist., *Meteorologica*, trans. Gerardi Cremonensis (books I–III) & Henrici Aristippi (book IV)
7. ***Libri de mineralibus*** [*De mineralibus*, Alberti Magni Opera omnia, Ed. Paris. v]	*corpus mixtum, in specie naturae constitutum per mixtionem, inanimatum, simplex*	Ps.-Arist., *De lapidibus* Avicenna, *De congelatione et conglutinatione lapidum*
8. ***Libri de anima*** [*De anima*, Alberti Magni Opera omnia, Ed. Colon. VII. 1]	*principium animatorum, i.e. ipsa anima secundum se et potentiae sive partes eius*	Arist., *De anima*, trans. vetus (Iacobus Venetus) & trans. Arabico-Latina (Mich. Scotus)
	opera et passiones animae in corpore:	
9. ***Liber de causa vitae et mortis et causis longioris vitae*** [*De morte et vita*, Alberti Magni Opera omnia, Ed. Paris. IX]	(1) *vita cui mors opponitur*	Arist., *De longitudine et brevitate vitae*, trans. vetus (Iacobus Venetus)
10. ***De nutrimento*** [*De nutrimento et nutrito*, Alberti Magni Opera omnia, Ed. Colon. VII. 2A]	(2) *nutrire*	Self-standing work of Albert; a reconstruction of Aristotle's lost work with the same title

Book [Edition]	**Subject Matter**	**Template / Main Source**
11. *Libri de somno et vigilia* [*De somno et vigilia*, Alberti Magni Opera omnia, Ed. Paris. IX]	(3) *opus sensibilis animae secundum comparationem sensus ad animal, sc. secundum quod sensus egreditur vel ingreditur in animal*	Arist., *De somno et vigilia, De insomniis, De divinatione per somnum*, trans. vetus (Anon.)
12. *Liber de sensu et sensato* [*De sensu et sensato*, Alberti Magni Opera omnia, Ed. Colon. VII. 2A]	(4) *opus sensibilis animae secundum comparationem sensus ad sensibile*	Arist., *De sensu et sensato*, trans. vetus (Anon.)
13. *Liber de memoria et reminiscentia* [*De memoria et reminiscentia*, Alberti Magni Opera omnia, Ed. Colon. VII. 2A]	(5) *opus sensibilis animae secundum reditum ex specie sensibili servata apud animam in rem prius acceptam in sensu*	Arist., *De memoria et reminiscentia*, trans. vetus (Iacobus Venetus)
14. *Liber de motibus animalium* [*De motibus animalium*, Alberti Magni Opera omnia, Ed. Paris. IX]	(6) *motus animae sensibilis secundum locum aut mutando locum aut dilatando et constringendo corpus in eodem loco*	Albert's reconstruction of the unavailable translation of Aristotle's *De motu animalium*
15. *De respiratione et inspiratione* [*De spiritu et respiratione*, Alberti Magni Opera omnia, Ed. Paris. IX]	(7) *specialis motus in habentibus pulmonem, qui est ad refrigerium pectoris per spiritum attractum, quem movet trahendo et retinendo et emitendo*	Costa ben Luca, *Liber de differentia spiritus et animae* [Albert no longer had the *vetus translatio* of Aristotle's *De respiratione* at his disposal]
16. *De intellectu et intelligibili* [Alberti Magni Opera omnia, Ed. Paris. IX]	(8) *opus animae secundum partem intellectualem*	Alexander de Aphr., *De intellectu et intellecto*; Alfarabi, *De intellectu et intellecto*; Alkindi, *De intellectu*; *Liber de causis*
17. *Libri de vegetabilibus* [*De vegetabilibus libri VII*, ed. by E. Meyer and C. Jessen, Berlin 1867]	*corpus animatum vegetabile*	Nicolaus Damascenus, *De plantis*, trans. Alfredi de Sareshel (template for books I–V); Avicenna, *Canon medicinae* (template for book VI); Palladius, *Opus agriculturae* (template for book VII)
18. *Libri de animalibus* [*De animalibus libri XXVI*, ed. by H. Stadler, Münster 1916–1920]	*corpus animatum sensibile quoad differentias animalium*	Arist., *Historia animalium, De generatione animalium, De partibus animalium*

It should be noted that this division of natural philosophy into different sciences (*scientiae naturales*) introduced and summarized Albert's very own system of the natural sciences. In the process of realizing his system, however, Albert added some works and slightly changed the order of this initial division, an order which was made on the basis of the order of things (*ordo rei*). In particular, he added three further works he had not mentioned in his *Physica*: *Liber de aetate sive de iuventute et senectute* (Alberti Magni Opera omnia, Ed. Paris. IX), *Liber de natura et origine animae* (Alberti Magni Opera omnia, Ed. Colon. XII), and *Liber de principiis motus processivi* (Alberti Magni Opera omnia, Ed. Colon. XII).[85] For reasons concerning an improved understanding of the scientific teachings contained in some of Albert's works, he eventually realized his system in the following order:

(1) *Physica* (including *De lineis indivisibilibus*)
(2) *De caelo et mundo*
(3) *De natura loci*
(4) *De causis proprietatum elementorum*
(5) *De generatione et corruptione*
(6) *Meteora*
(7) *De mineralibus*
(8) *De anima*
(9) *De nutrimento et nutrito*
(10) *De sensu et sensato*
(11) *De memoria et reminiscentia*
(12) *De intellectu et intelligibili*
(13) *De somno et vigilia*
(14) *De spiritu et respiratione*
(15) *De motibus animalium*
(16) *Liber de aetate sive de iuventute et senectute*
(17) *De morte et vita*
(19) *De vegetabilibus*
(20) *De animalibus*
(21) *Liber de natura et origine animae*
(22) *Liber de principiis motus processivi*

85 Concerning this point, see also Anzulewicz, 'Anhang zu *Physica* l. 1 tr. 1 c. 4.'.

Appendix II: Overview of the Sources of Albert's *Physica*[86]

(Italicized author names indicate that Albert cites the author indirectly, through another source.)

Explicit Sources	**Implicit Sources (determined by Hossfeld)**
Abubacher *alias* Habonycer = Abū Bakr	Al-Bitrūjī
al-Rāzī	*De motibus caelorum*
Academici	Alexander Halensis
Alboali = Avicenna[87]	*Glossa in quatuor libros*
Alcmaeon Crotoniates	*Sententiarum*
Alexander de Aphrodisias (A. Graecus,	*Summa fratris Alexandri*
A. Peripateticus)	Algazel
Alfarabi (Alfarabius)	*Metaphysica*
De ortu scientiarum	Al-Ǧūzǧānī
Liber de scientiis	*Praefatio in Librum Sufficientiae*
Algazel	*Avicennae*[88]
Metaphysica	Anselmus Cantuariensis
Anaxagoras	*Monologion*
Apuleius (Ps.-)	*Proslogion*
Asclepius	Apuleius
Aristoteles	*De mundo*
Analytica posteriora	Apuleius (Ps.-)
Analytica priora	*Asclepius*
De anima	Aristoteles
De caelo et mundo	*Analytica priora*
Categoriae	*Analytica posteriora*
Ethica Nicomachea	*De anima*
De generatione animalium	*De caelo et mundo*
De generatione et corruptione	*Categoriae*
Metaphysica	*De divinatione per somnum*
Meteora	*Ethica Nicomachea*
Physica	*De generatione animalium*
De sensu et sensato	*De generatione et corruptione*
De somno et vigilia	*De historia animalium*
Topica	*Metaphysica*
Aristoteles (Ps.-)	*Meteora*
De natura deorum	*De motu animalium*
Augustinus	*De partibus animalium*
Confessiones	*Physica*
Avempace	*De sensu et sensato*

86 Both lists were made on the basis of the source indices *Auctores ab Alberto ipso allegati* and *Auctores a nobis allegati* of the Ed. Colon. IV. 1–2.

87 See Bertolacci, 'Albert the Great', esp. p. 151, note 59.

88 See Bertolacci, 'Albert the Great'; Birkenmajer, 'Avicennas Vorrede zum "Liber Sufficientiae"'.

Explicit Sources	**Implicit Sources (determined by Hossfeld)**
Averroes	*De somno et vigilia*
De anima	*Topica*
Metaphysica	Aristoteles (Ps.-)
Physica	*De proprietatibus elementorum*
Avicebron	*De natura deorum*
Fons vitae	Augustinus
Avicenna	*Confessiones*
De generatione et corruptione	Augustinus (Ps.-)
Liber de anima (VI De naturalibus)	*Paraphrasis Themistiana*
Philosophia prima	Averroes
Sufficientia	*De anima*
Boethius	*De caelo et mundo*
De consolatione philosophiae	*De generatione et corruptione*
Contra Eutychen et Nestorium	*Metaphysica*
De institutione arithmetica	*Physica*
In Isagogen Porphyrii Ed. II	Avicebron
In Peri hermeneias Ed. II	*Fons vitae*
De trinitate	Avicenna
Brisso	*De animalibus*
Chalcidius	*De generatione et corruptione*
Commentarius in Timaeum Platonis	*Liber de anima (VI De naturalibus)*
Costa ben Luca	*Philosophia prima*
De differentia animae et spiritus	*Sufficientia*
David de Dinant	Boethius
Democritus	*De consolatione philosophiae*
Dionysius Areopagita (Ps.-)	*Contra Eutychen et Nestorium*
De divinis nominibus	*Geometria*
Empedocles	*De institutione arithmetica*
Epicurus et Epicurei	*De institutione musica*
Euclides	*In Isagogen Porphyrii Ed. II*
Elementa	*In Peri hermeneias Ed. II*
Eustratius	*De trinitate*
In Ethicam Nicomacheam	Bonaventura
Firmicius Maternus	*In II librum Sententiarum*
Mathesis	Chalcidius
Gal(i)enus	*Commentarius in Timaeum Platonis*
Gal(i)enus Arabus	Cicero
Gilbertus Porretanus	*De natura deorum*
Commentarii in Boethii librum de trinitate	Costa ben Luca
Gilbertus Porretanus (Ps.-)	*De differentia animae et spiritus*
Liber sex principiorum	Dionysius Areopagita (Ps.-)
Grammatici	*De divinis nominibus*
Habonycer Maurus *vide* Abubacher	*De mystica theologia*
Heraclitus	Euclides
Hermes Trismegistus	*Elementa*
Hesiodistae	Gilbertus Porretanus
	Commentarii in Boethii librum de trinitate

Explicit Sources	**Implicit Sources (determined by Hossfeld)**
Homerus	Gilbertus Porretanus (Ps.-)
Iohannes Damascenus	*Liber sex principiorum*
De fide orthodoxa	Iohannes Damascenus
Iohannes Grammaticus (Philoponos)	*De fide orthodoxa*
Isaac Israeli	Isaac Israeli
Liber de definicionibus	*Liber de definicionibus*
Leucippus	Isidorus Hispalensis
Liber de causis	*Eytmologiae*
Machometus	*Liber de causis*
Mathematici	Martianus Capella
Melissus	*De nuptiis Philologiae et Mercurii*
Messahallach	Moses Maimonides
De coniunctionibus planetarum	*Dux neutrorum*
Moses Maimonides (M. Aegyptius)	Nicolaus Damascenus
Ovidius	*De plantis*
Metamorphoses	Ovidius
Parmenides	*Metamorphoses*
Peripatetici	*Remedia amoris*
Philosophi	Papias Vocabulista
Physici	*Elementarium doctrine rudimentum*
Plato	Petrus Abaelardus
Phaedrus	*Logica 'Ingredientibus'*
Timaeus	Petrus Hispanus
Platonici	*De fallaciis*
Poetae	Petrus Lombardus
Porphyrius	*Liber I Sententiarum*
Isagoge	Porphyrius
Ptolemaeus	*Isagoge*
Almagestum	Seneca
Liber Quadripartiti	*Naturales quaestiones*
Ptolemaeus (Ps.-)	Simon Aurea Capra
Centiloquium	*Ilias*
Pythagoras	Vergilius
Pythagorici	*Aenaeis*
Quidam	Vitruvius
Sapientes	*De architectura*
Seneca	
Naturales quaestiones	
Simon Aurea Capra	
Ilias	
Socrates	
Sophistae	
Speusippus	
Stoici	
Themistius	
Theologi	
Theoprastus	
Vergilius	
Xutos	
Zenon	

Bibliography

Manuscripts

Albertus Magnus, 'Politica', Berlin, Staatsbibliothek Preußischer Kulturbesitz, MS lat. 879

John Damascene, 'De fide ortodoxa, libri 1–4. Interpretatio latina Iohannis Burgundionis iudicis Pisani', Città della Vaticano, Biblioteca Apostolica Vaticana, MS Pal. lat. 314

Primary Sources

Albertus Magnus, *Analytica posteriora*, ed. by A. Borgnet, Alberti Magni Opera Omnia, Ed. Paris. II (Paris: Vivès, 1890), pp. 1–232

———, *De intellectu et intelligibili*, ed. by A. Borgnet, Alberti Magni Opera Omnia, Ed. Paris. IX (Paris: Vivès, 1890), pp. 477–521

———, *De sophisticis elenchis*, ed. by A. Borgnet, Alberti Magni Opera Omnia, Ed. Paris. IX (Paris: Vivès, 1890), pp. 525–713

———, *Politica*, ed. by A. Borgnet, Alberti Magni Opera Omnia, Ed. Paris. VIII (Paris: Vivès, 1891)

———, *Ethica*, ed. by A. Borgnet, Alberti Magni Opera Omnia, Ed. Paris. VII (Paris: Vivès, 1891)

———, *De IV coaequaevis*, ed. by A. Borgnet, Alberti Magni Opera Omnia, Ed. Paris. XXXIV (Paris: Vivès, 1895), pp. 307–761

———, *De animalibus libri XXVI. Nach der Cölner Urschrift*, ed. by H. Stadler, Beiträge zur Geschichte der Philosophie des Mittelalters, XV (Münster: Aschendorff, 1916–1920)

———, *De bono*, ed. by H. Kühle, C. Feckes, B. Geyer, and W. Kübel, Alberti Magni Opera Omnia, Ed. Colon. XXVIII (Münster: Aschendorff, 1951)

———, *Liber de natura et origine animae*, ed. by B. Geyer, Alberti Magni Opera Omnia, Ed. Colon. XII (Münster: Aschendorff, 1955), pp. 1–44

———, *Metaphysica, Libri I–V*, ed. by B. Geyer, Alberti Magni Opera Omnia, Ed. Colon. XVI. 1 (Münster: Aschendorff, 1960)

———, *Metaphysica, Libri VI–XIII*, ed. by B. Geyer, Alberti Magni Opera Omnia, Ed. Colon. XVI. 2 (Münster: Aschendorff, 1964)

———, *De anima*, ed. by C. Stroick, Alberti Magni Opera Omnia, Ed. Colon. VII. 1 (Münster: Aschendorff, 1968)

———, *Super Ethica. Commentum et quaestiones, libri I–V*, ed. by W. Kübel, Alberti Magni Opera Omnia, Ed. Colon. XIV. 1 (Münster, 1968)

———, *Super Dionysium De divinis nominibus*, ed. by P. Simon, Alberti Magni Opera Omnia, Ed. Colon. XXXVII. 1 (Münster: Aschendorff, 1972)

———, *De natura boni*, ed. by E. Filthaut, Alberti Magni Opera Omnia, Ed. Colon. XXV. 1 (Münster: Aschendorff, 1974)

———, *Summa theologiae sive de mirabili dei, libri I pars 1, quaest. 1-50A*, ed. by D. Siedler, W. Kübel, and H.-G. Vogels, Alberti Magni Opera Omnia, Ed. Colon. XXXIV. 1 (Münster: Aschendorff, 1978)

———, *Super Dionysii Epistulas*, ed. by P. Simon, Alberti Magni Opera Omnia, Ed. Colon. XXXVII. 2 (Münster: Aschendorff, 1978), pp. 479–554
———, *Physica*, libri 1–4, ed. by P. Hossfeld, Alberti Magni Opera Omnia, Ed. Colon. IV. 1 (Münster: Aschendorff, 1987)
———, *Super Matthaeum*, ed. by B. Schmidt, Alberti Magni Opera Omnia, Ed. Colon. XXI. 1–2 (Münster: Aschendorff, 1987)
———, *De causis et processu universitatis a prima causa*, ed. by W. Fauser, Alberti Magni Opera Omnia, Ed. Colon. XVII. 2 (Münster: Aschendorff, 1993)
———, *Quaestio de primis motibus*, ed. by A. Fries, W. Kübel, and H. Anzulewicz, Alberti Magni Opera Omnia, Ed. Colon. XXV. 2 (Münster: Aschendorff, 1993), pp. 1–12
———, *Quaestio de synderesi*, ed. by A. Fries, W. Kübel, and H. Anzulewicz, Alberti Magni Opera Omnia, Ed. Colon. XXV. 2 (Münster: Aschendorff, 1993), pp. 232–38
———, *Super Dionysium De caelesti hierarchia*, ed. by P. Simon and W. Kübel, Alberti Magni Opera Omnia, Ed. Colon. XXXVI. 1 (Münster: Aschendorff, 1993)
———, *Meteora*, ed. by P. Hossfeld, Alberti Magni Opera Omnia, Ed. Colon. VI. 1 (Münster: Aschendorff, 2003)
———, *Super Porphyrium De V universalibus*, ed. by M. Santos Noya, Alberti Magni Opera Omnia, Ed. Colon. I. 1A (Münster: Aschendorff, 2004)
———, *Liber divisionum*, ed. by H. Möhle, Alberti Magni Opera Omnia, Ed. Colon. I. 2 (Münster: Aschendorff, 2006), pp. 81–125
———, *De homine*, ed. by H. Anzulewicz and J. R. Söder, Alberti Magni Opera Omnia, Ed. Colon. XVII. 2 (Münster: Aschendorff, 2008)
———, *De praedicamentis*, ed. by M. Santos Noya, C. Steel, and S. Donati, Alberti Magni Opera Omnia, Ed. Colon. I. 1B (Münster: Aschendorff, 2013)
———, *Super I librum Sententiarum. Distinctiones 1–3*, ed. by M. Burger, Alberti Magni Opera Omnia, Ed. Colon. XXIX. 1 (Münster: Aschendorff, 2015)
Alexander Halensis, *Summa theologica, II, 1*, ed. by PP. Collegii S. Bonaventurae(Quaracchi: Ex Typographia Collegii S. Bonaventurae, 1928)
Algazel's Metaphysics: A Mediaeval Translation, ed. by J. T. Muckle (Toronto: PIMS, 1933)
'Logica Algazelis Introduction and Critical Texts', ed. by C. H. Lohr, *Traditio*, 21 (1965), 223–90
Aristoteles Graece. Ex recensione Immanuelis Bekkeri (Berlin: Academia Regia Borussica, 1831)
Avicenna, *Logica* (Venezia: O. Scotus, 1508, repr. Frankfurt am Main: Minerva, 1961)
———, *Liber de Philosophia prima sive Scientia divina I–IV*, ed. by S. Van Riet (Louvain: Peeters, 1977)
Guilielmus Alvernus, *Tractatus De anima*, Guilielmi Alverni Opera Omnia, II. 2 (Paris: A. Pralard, 1674), pp. 65–228
Némésius d'Émèse, *De natura hominis*, trans. by Burdundio de Pisa, ed. by G. Verbeke and J. R. Moncho, Corpus Latinum Commentariorum in Aristotelem Graecorum, suppl. 1 (Leiden: Brill, 1975)

Philippus Cancellarius, *Summa de bono*, ed. by N. Wicki, Corpus Philosophorum Medii Aevi, Opera Philosophica Mediae Aetatis selecta 11 (Bern: Francke, 1985)

Rufini Aquileiensis Homiliarum Basilii Magni interpretatio Latina, ed. by C. Lo Cicero, Corpus Christianorum Series Latina, 20A (Turnhout: Brepols, 2008)

Secondary Studies

Anzulewicz, Henryk, *Die theologische Relevanz des Bildbegriffs und des Spiegelbildmodells in den Frühwerken des Albertus Magnus*, Beiträge zur Geschichte der Philosophie und Theologie des Mittelalters N.F. 53, 1–2 (Münster: Aschendorff, 1999)

———, 'Entwicklung und Stellung der Intellekttheorie im System des Albertus Magnus', *Archives d'histoire doctrinale et littéraire du Moyen Âge*, 70 (2003), 165–218

———, 'Vermögenspsychologische Grundlagen kognitiver Leistung des Intellektes nach Albertus Magnus', *Acta Mediaevalia*, 22 (2009), 93–116

———, 'Solus homo est nexus Dei et mundi. Albertus Magnus über den Menschen', in *Multifariam: homenaje a los profesores Anneliese Meis, Antonio Bentué y Sergio Silva*, ed. by S. Fernández, J. Noemi, and R. Polanco. Anales de la Facultad de Teología, 1 (Santiago de Chile: Facultad de Teología, 2010), pp. 321–35

———, 'Alberts Konzept der Bildung durch Wissenschaft', in *Albertus Magnus und der Ursprung der Universitätsidee: die Begegnung der Wissenschaftskulturen im 13. Jahrhundert und die Entdeckung des Konzepts der Bildung durch Wissenschaft*, ed. by L. Honnefelder (Berlin: Berlin University Press, 2011), pp. 382–97, 538–48

———, 'Anhang zu *Physica* l. 1 tr. 1 c. 4. Alberts *libri naturales* und Lehrsätze, die von Aristoteles abweichen', in *Albertus Magnus und sein System der Wissenschaften: Schlüsseltexte in Übersetzung Lateinisch – Deutsch*, ed. by Albertus-Magnus-Institut (Münster: Aschendorff, 2011), pp. 155–59

———, 'Anthropology: The Concept of Man in Albert the Great', in *A Companion to Albert the Great: Theology, Philosophy, and the Sciences*, ed. by I. M. Resnick, Brill's Companions to the Christian Tradition, 38 (Leiden: Brill, 2012), pp. 325–46

———, 'Albertus Magnus und seine Schüler. Versuch einer Verhältnisbestimmung', in *Schüler und Meister*, ed. by A. Speer, Miscellanea Mediaevalia, 39 (Berlin: de Gruyter, 2016), pp. 159–203

Anzulewicz, Henryk, and Katja Krause, 'Albert der Große und sein holistisches Konzept menschlicher Erkenntnis', in *Veritas et subtilitas: Truth and Subtlety in the History of Philosophy: Essays in memory of Burkhard Mojsisch (1944–2015)*, ed. by T. Iremadze and U. R. Jeck, Bochumer Studien zur Philosophie, 59 (Amsterdam: John Benjamins, 2018), pp. 157–94

Anzulewicz, Henryk, and Caterina Rigo, 'Reductio ad esse divinum. Zur Vollendung des Menschen nach Albertus Magnus', in *Ende und Vollendung: Eschatologische Perspektiven im Mittelalter*, ed. by J. A. Aertsen, Miscellanea Mediaevalia, 29 (Berlin: de Gruyter, 2002), pp. 388–416

Bertolacci, Amos, 'La divisione della filosofia nel primo capitolo del Commento di Alberto Magno alla Fisica: Le fonti Avicenniane', in *La Divisione della Filosofia e le sue Ragioni. Lettura di testi medievali* (VI–XIII sec.), ed. by G. d'Onofrio (Cava de' Tirreni: Avvagliano, 2001), pp. 137–55

———, 'Albert the Great and the Preface of Avicenna's Kitāb Al-Šifā', in *Avicenna and his Heritage: Acts of the International Colloquium, Leuven–Louvain-la-Neuve, September 8–September 11, 1999*, ed. by J. L. Janssens and D. De Smet, Ancient and Medieval Philosophy, 28 (Leuven: Leuven University Press, 2002), pp. 131–53

Birkenmajer, Aleksander, 'Avicennas Vorrede zum "Liber Sufficientiae" und Roger Bacon', in *Études d'Histoire des Sciences et de la Philosophie du Moyen Âge*, ed. by A. Birkenmajer, Studia Copernicana, 1 (Warsaw: Ossolineum, 1970), pp. 89–101

Brentjes, Sonja, 'Reflexionen zur Bedeutung der im 12. Jahrhundert angefertigten lateinischen Übersetzungen wissenschaftlicher Texte für die europäische Wissenschaftsgeschichte', in *Europa: die Gegenwärtigkeit der antiken Überlieferung*, ed. by J. Cobet, Essener Beiträge zur Kulturgeschichte, 2 (Aachen: Shaker, 2000), pp. 269–305

Brentjes, Sonja, Alexander Fidora, and Matthias M. Tischler, 'Towards a New Approach to Medieval Cross-Cultural Exchanges', *Journal of Transcultural Medieval Studies*, 1 (2014), 9–50

Brumberg-Chaumont, Julie, 'Les divisions de la logique selon Albert le Grand', in *Ad notitiam ignoti. L'Organon dans la translatio studiorum à l'époque d'Albert le Grand*, ed. by J. Brumberg-Chaumont, Studia Artistarum: Études sur la Faculté des arts dans les Universités médiévales, 37 (Turnhout: Brepols, 2013), pp. 335–416

Burger, Maria, 'Albertus Magnus. Theologie als Wissenschaft unter der Herausforderung aristotelisch-arabischer Wissenschaftstheorie', in *Albertus Magnus und der Ursprung der Universitätsidee: Die Begegnung der Wissenschaftskulturen im 13. Jahrhundert und die Entdeckung des Konzepts der Bildung durch Wissenschaft*, ed. by L. Honnefelder (Berlin: Berlin University Press, 2011), pp. 97–114, 451–63

Burnett, Charles, 'The Institutional Context of Arabic-Latin Translations of the Middle Ages. A Reassessment of the "School of Toledo"', in *Vocabulary of Teaching and Research between Middle Ages and Renaissance: Proceedings of the Colloquium, London, Warburg Institute, 11–12 March 1994*, ed. by O. Weijers, Etudes sur le vocabulaire intellectuel du Moyen Âge, Comité International du Vocabulaire des Institutions et de la Communication Intellectuelles au Moyen Âge (CIVICIMA), 8 (Turnhout: Brepols, 1995), pp. 214–35

———, 'Translating from Arabic into Latin in the Middle Ages: Theory, Practice, and Criticism', in *Éditer, traduire, interpréter: essais de méthodologie philosophique*, ed. by P. W. Rosemann and S. G. Lofts, Philosophes médiévaux, 36 (Louvain-la-Neuve: Peeters, 1997), pp. 55–78

———, *Arabic into Latin in the Middle Ages: The Translators and their Intellectual and Social Context*, Variorum Collected Studies (Farnham: Ashgate, 2009)

Chung, Hyun-Suk [Stephen Chung], 'Dissociating Angels and Intelligences: A Medieval Angelological Problem and Albert the Great', *Sogang Journal of Philosophy*, 39 (2014), 211–38

D'Alverny, Marie-Thérèse, 'Les traductions à deux interprètes: d'arabe en langue vernaculaire et de langue vernaculaire en latin', in *Traduction et traducteurs au Moyen Âge: Actes du colloque international du CNRS organisé à Paris, Institut de Recherche et d'Histoire des Textes les 26–28 mai 1986*, ed. by G. Contamine, Documents, études et repertoires (Paris: CNRS, 1989), pp. 193–206

Davidson, Herbert A., *Alfarabi, Avicenna, and Averroes, on Intellect: Their Cosmologies, Theories of the Active Intellect, and Theories of Human Intellect* (New York: Oxford University Press, 1992)

de Asúa, Miguel, 'Minerals, Plants and Animals from A to Z: The Inventory of the Natural Word in Albert the Great's *philosophia naturalis*', in *Albertus Magnus: zum Gedenken nach 800 Jahren: Neue Zugänge, Aspekte und Perspektiven*, ed. by W. Senner and H. Anzulewicz, Quellen und Forschungen zur Geschichte des Dominikanerordens, 10 (Berlin: Akademie, 2001), pp. 389–400

———, 'War and Peace. Medicine and Natural Philosophy in Albert the Great', in *A Companion to Albert the Great: Theology, Philosophy, and the Sciences*, ed. by I. M. Resnick, Brill's Companions to the Christian Tradition, 38 (Leiden: Brill, 2013), pp. 269–97

de Leemans, Pieter, 'Aristotle Transmitted: Reflections on the Transmission of Aristotelian Scientific Thought in the Middle Ages', *International Journal of the Classical Tradition*, 17 (2010), 325–53

de Smet, Daniel, '"Perfectio prima—perfectio secunda" ou les vicissitudes d'une notion: de S. Thomas aux Ismaéliens Ṭayyibites du Yémen', *Recherches de théologie et philosophie médiévales*, 66 (1999), 254–88

Dodd, Bernard G., 'Aristoteles Latinus', in *The Cambridge History of Later Medieval Philosophy: From the Rediscovery of Aristotle to the Disintegration of Scholasticism, 1100–1600*, ed. by N. Kretzmann and A. Kenny (Cambridge: Cambridge University Press, 1997), pp. 45–79

Donati, Silvia, 'Alberts des Großen Konzept der *scientiae naturales*: Zur Konstitution einer peripatetischen Enzyklopädie der Naturwissenschaften', in *Albertus Magnus und der Ursprung der Universitätsidee: Die Begegnung der Wissenschaftskulturen im 13. Jahrhundert und die Entdeckung des Konzepts der Bildung durch Wissenschaft*, ed. by L. Honnefelder (Berlin: Berlin University Press, 2011), pp. 354–81, 524–38

Falcon, Andrea, 'Commentators on Aristotle', *The Stanford Encyclopedia of Philosophy Archive, Fall 2017 Edition*, ed. by E. N. Zalta, <https://plato.stanford.edu/archives/fall2017/entries/aristotle-commentators/> [accessed 28 January 2020]

Freudenthal, Gad, 'Arabic and Latin Cultures as Resources for the Hebrew Translation Movement, Comparative Considerations, both Quantitative and Qualitative', in *Science in Medieval Jewish Cultures*, ed. by G. Freudenthal (Cambridge: Cambridge University Press, 2011), pp. 74–104

Goyens, Michèle, Pieter de Leemans, and An Smets, eds, *Science Translated: Latin and Vernacular Translations of Scientific Treatises in Medieval Europe*, Mediaevalia Lovaniensia, 40 (Leuven: Leuven University Press, 2008)

Grabmann, Martin, 'Der Einfluss Alberts des Grossen auf das mittelalterliche Leben', in *Mittelalterliches Geistesleben. Abhandlungen zur Geschichte der Scholastik und Mystik*, vol. 2, ed. by M. Grabmann (Munich: Max Hueber, 1936), pp. 324–412

Günsz, Aron, *Die Abhandlung Alexanders von Aphrodisias über den Intellekt. Aus handschriftlichen Quellen zum ersten Male herausgegeben und durch die Abhandlung: 'Die Nûslehre Alexanders von Aphrodisias und ihr Einfluss auf die arabisch-jüdische Philosophie des Mittelalters' eingeleitet* (Berlin: Itzkowski, 1886)

Hasse, Dag N., 'Das Lehrstück von den Vier Intellekten in der Scholastik—von den Arabischen Quellen bis zu Albertus Magnus', *Recherches de théologie et philosophie médiévales*, 66 (1999), 21–78

Hasselhoff, Görge K., 'Hieronymus und die Aussprache des Hebräischen nach seinen Briefen', *Jahrbuch für Antike und Christentum*, 59 (2016), 43–66

Honnefelder, Ludger, '"Bildung durch Wissenschaft": Albertus Magnus oder die Geburt der Universitätsidee aus der Begegnung griechisch-hebräisch-arabischer Wissenschaft mit christlicher Gelehrsamkeit im lateinischen Westen des 13. Jahrhunderts. Eine Einführung', in *Albertus Magnus und der Ursprung der Universitätsidee: Die Begegnung der Wissenschaftskulturen im 13. Jahrhundert und die Entdeckung des Konzepts der Bildung durch Wissenschaft*, ed. by L. Honnefelder (Berlin: Berlin University Press, 2011), pp. 9–23

———, ed., *Albertus Magnus und der Ursprung der Universitätsidee: Die Begegnung der Wissenschaftskulturen im 13. Jahrhundert und die Entdeckung des Konzepts der Bildung durch Wissenschaft* (Berlin: Berlin University Press, 2011)

———, 'Metaphysik als "Erste Wissenschaft": Die kritische Rezeption der aristotelischen Metaphysik durch Albert den Großen', in *Albertus Magnus und der Ursprung der Universitätsidee: Die Begegnung der Wissenschaftskulturen im 13. Jahrhundert und die Entdeckung des Konzepts der Bildung durch Wissenschaft*, ed. by L. Honnefelder (Berlin: Berlin University Press, 2011), pp. 332–53

———, *Albertus Magnus und die kulturelle Wende im 13. Jahrhundert—Perspektiven auf die epochale Bedeutung des großen Philosophen und Theologen*, Lectio Albertina, 13 (Münster: Aschendorff, 2012)

Hornig, Gottfried, 'Perfektibilität: Eine Untersuchung zur Geschichte und Bedeutung dieses Begriffs in der deutschsprachigen Literatur', *Archiv für Begriffsgeschichte*, 24 (1980), 221–57

Jordan, Mark D., 'Albert the Great and the Hierarchy of Sciences', *Faith and Philosophy: The Journal of the Society of Christian Philosophers*, 9 (1992), 483–99

Jüssen, Gabriel, 'Wilhelm von Auvergne und die Transformation der scholastischen Philosophie im 13. Jahrhundert', in *Philosophie im Mittelalter: Entwicklungslinien und Paradigmen*, ed. by J. P. Beckmann and W. Kluxen (Hamburg: Meiner, 1996), pp. 141–64

Köhler, Theodor W., *Grundlagen des philosophisch-anthropologischen Diskurses im dreizehnten Jahrhundert: Die Erkenntnisbemühung um den Menschen im zeitgenössischen Verständnis*, Studien und Texte zur Geistesgeschichte des Mittelalters, 71 (Leiden: Brill, 2000)

Köpf, Ulrich, and Dieter R. Bauer, eds, *Kulturkontakte und Rezeptionsvorgänge in der Theologie des 12. und 13. Jahrhunderts*, Archa Verbi. Subsidia, 8 (Münster: Aschendorff, 2011)

Krause, Katja, 'Grenzen der Philosophie: Alberts des Großen Kommentar zu *De animalibus* und die Medizin', *Documenti e studi sulla tradizione filosofica medievale*, 30 (2019), 265–93

Layer, Adolf, 'Namen und Ehrennamen Alberts des Großen', in *Albert von Lauingen: 700 Jahre † Albertus Magnus*, Festschrift 1980 (Lauingen: Historischer Verein Dillingen, 1980), pp. 41–43

Lefevere, André, ed., *Translation, History, Culture: A Sourcebook*, Translation Studies (London: Taylor & Francis eBooks, 2003)

Möhle, Hannes, 'Albertus Magnus und die Vielfalt der Wissenschaften', in *Albertus Magnus und der Ursprung der Universitätsidee: Die Begegnung der Wissenschaftskulturen im 13. Jahrhundert und die Entdeckung des Konzepts der Bildung durch Wissenschaft*, ed. by L. Honnefelder (Berlin: Berlin University Press, 2011), pp. 301–31, 511–20

Moraux, Paul, *Alexandre d'Aphrodise Exégète de la Noétique d'Aristote* (Paris: Belles Lettres, 1942)

Müller, Jörn, *Natürliche Moral und philosophische Ethik bei Albertus Magnus*, Beiträge zur Geschichte der Philosophie und Theologie des Mittelalters, NF, 59 (Münster: Aschendorff, 2001)

Robertson, Duncan, *Lectio Divina: The Medieval Experience of Reading*, Cistercian Studies, 238 (Collegeville: Liturgical Press, 2011)

Rodolfi, Anna, '"Ex nihilo id est post nihilum". Alberto Magno e il dibattito sull'eternità del mondo', *Studi medievali*, 40 (1999), 681–704

———, 'La dottrina della materia celeste in Alberto Magno', *Medioevo e rinascimento*, 14 (2003), 25–48

———, *Il concetto di materia nell'opera di Alberto Magno*, Corpus Philosophorum Medii Aevi. Testi e Studi, 18 (Firenze: SISMEL Edizioni del Galluzzo, 2004)

Roling, Bernd, *Locutio angelica: die Diskussion der Engelsprache als Antizipation einer Sprechakttheorie in Mittelalter und Früher Neuzeit*, Studien und Texte zur Geistesgeschichte des Mittelalters, 97 (Leiden: Brill, 2008)

Rosier-Catach, Irène, 'The Augustinian Threefold Word and Angelic Mental Speech' in *The Word in Medieval Logic, Theology and Psychology: Acts of the XIIIth International Colloquium of the Société Internationale pour l'Étude de la Philosophie Médiévale, Kyoto, 27 September–1 October 2005*, ed. by T. Shimizu and C. Burnett, Rencontres de Philosophie Médiévale, 14 (Turnhout: Brepols, 2009), pp. 16–48

Sabra, Abdelhamid I., 'The Appropriation and Subsequent Naturalization of Greek Science in Medieval Islam: A Preliminary Statement', in *Tradition, Transmission, Transformation: Proceedings of Two Conferences on Pre-Modern Science Held at the University of Oklahoma*, ed. by F. J. Ragep, S. P. Ragep, and S. Livesey, Collection de Travaux de l'Académie Internationale d'Histoire des Sciences, 37 (Leiden: Brill, 1996), pp. 3–27

Schipperges, Heinrich, 'Das medizinische Denken bei Albertus Magnus', in *Albertus Magnus – Doctor universalis: 1280/1980*, ed. by G. Meyer, A. Zimmermann, and P.-B. Lüttringhaus, Walberberger Studien, Philosophische Reihe, 6 (Mainz: Matthias Grünewald, 1980), pp. 279–94

———, 'Eine "summa medicinae" bei Albertus Magnus', in *Jahres- und Tagungsbericht der Görres-Gesellschaft 1980* (Aachen: Görres-Gesellschaft, 1981), pp. 5–24

Schneider, Arthur C. A., *Die Psychologie des Albertus Magnus. 1. Teil*, Beiträge zur Geschichte der Philosophie des Mittelalters, 5 (Münster: Aschendorff, 1903)

Siraisi, Nancy G., 'The Medical Learning of Albertus Magnus', in *Albertus Magnus and the Sciences: Commemorative Essays 1980*, ed. by J. A. Weisheipl, Studies and Texts, 49 (Toronto: Pontifical Institute of Mediaeval Studies, 1980), pp. 379–404

Speer, Andreas, and Lydia Wegener, eds, *Wissen über Grenzen: Arabisches Wissen und lateinisches Mittelalter*, Miscellanea Mediaevalia, 33 (Berlin: de Gruyter, 2006)

Steel, Carlos, *Der Adler und die Nachteule: Thomas und Albert über die Möglichkeit der Metaphysik*, Lectio Albertina, 4 (Münster: Aschendorff, 2001)

Sturlese, Loris, *Vernunft und Glück: Die Lehre vom 'intellectus adeptus' und die mentale Glückseligkeit bei Albert dem Großen*, Lectio Albertina, 7 (Münster: Aschendorff, 2005)

Tremblay, Bruno, 'Albert le Grand: De ce qui vient avant la logique', *History and Philosophy of Logic*, 25 (2004), 165–203

Trizio, Michele, 'Greek Aristotelian Works Translated into Latin', in *The Cambridge History of Medieval Philosophy*, vol. 1–2, ed. by R. Pasnau and C. van Dyke (Cambridge: Cambridge University Press, 2010), pp. 793–97

Twetten, David B., 'Albert the Great's Early Conflations of Philosophy and Theology on the Issue of Universal Causality', in *Medieval Masters: Essays in Memory of Msgr. E. A. Synan*, ed. by R. E. Houser and E. A. Synan, Thomistic Papers, 7 (Houston: Center for Thomistic Studies, 1999), pp. 25–63

Vosté, Jacob-Maria, 'S. Albertus Magnus Evangeliorum Interpres', *Angelicum*, 9 (1932), 239–327

———, 'Sanctus Albertus Magnus Prophetarum Interpres', *Divus Thomas*, 9 (1932), 469–506

Winkler, Norbert, 'Seele–Engel–Intelligenzen. Ein philosophiegeschichtlicher Einblick', in *Fragmenta Melanchthoniana*, vol. 3: *Melanchthons Wirkung in der europäischen Bildungsgeschichte*, ed. by G. Frank and S. Lalla (Heidelberg: Verlag Regionalkultur, 2007), pp. 239–64

Wisnovsky, Robert, Faith Wallis, Jamie C. Fumo, and Carlos Fraenkel, 'Introduction: Vehicles of Transmission, Translation, and Transformation in Medieval Textural Culture', in *Vehicles of Transmission, Translation, and Transformation in Medieval Textual Culture*, Cursor Mundi, 4 (Turnhout: Brepols, 2011), pp. 1–22

Zonta, Mauro, *Hebrew Scholasticism in the Fifteenth Century: A History and Source Book*, Amsterdam Studies in Jewish Thought, 9 (Dordrecht: Springer, 2006)

CLAIRE GILBERT

Arabic from the Margins

Hispano-Moroccan Translation between Classical Arabic and Humanist Traditions in Early Modern Spain

Introduction: Arabic in Sixteenth-Century Spain

In the sixteenth century, the Castilian vernacular was codified domestically and abroad through publications like Nebrija's *Gramática castellana* (Salamanca, 1492) or foreign Spanish dictionaries like the *Vocabulario para aprender frances, español, y flamenco* (Antwerp, 1527) or *The Spanish Schoole Master* (London, 1591). This trend for producing vernacular lexicographical tools, in print, for an expanding market of language learners is one of the hallmarks of the history of language and literacy in early modern Europe.[1] In the same period, another Spanish language — Arabic, in its classical and vernacular forms — was suffering a long-lasting process of legal and demographic erasure in Europe. This process began with legal proscriptions against using Arabic or possessing Arabic texts which were first promulgated between 1500 and 1526 and renewed or revived throughout the century, and culminated with the forced expulsion of Spain's *moriscos*, a decision justified in large part using arguments about language and ethnicity.[2] Nevertheless, Spanish subjects and others continued to use, and learn, Arabic in Spain throughout the sixteenth century.[3] Spanish Arabism served religious, political, and economic needs.

1 Acero Durántez, 'Lexicografía plurilingüe de los siglos XVI y XVII'; Houston, *Literacy in Early Modern Europe Culture and Education.*

2 García-Arenal, 'The Religious Identity of the Arabic Language'; Feros, 'Rhetorics of Expulsion'.

3 García-Arenal and Rodríguez Mediano, 'Sacred History, Sacred Languages'; Martínez de Castilla Muñoz, 'The Teaching and Learning of Arabic in Salamanca'.

Claire Gilbert • (claire.gilbert@slu.edu) is an associate professor of Early Modern History at Saint Louis University. She is the author of *In Good Faith: Arabic Translation and Translators in Early Modern Spain* (Philadelphia: University of Pennsylvania Press, 2020) as well as articles and book chapters about the history of Arabic and Spanish translation in the Western Mediterranean.

Premodern Translation: Comparative Approaches to Cross-Cultural Transformations, ed. by Sonja Brentjes and Alexander Fidora, Contact and Transmission, 2 (Turnhout: Brepols, 2021), pp. 133–164

10.1484/M.CAT-EB.5.122136

Indeed, although scholarly interest in early modern Spanish Arabism is perhaps reaching its apogee, one aspect of the study of this Arabism which has not received sufficient analysis is the heterogeneity of the subject matter. Put simply: not all Arabism in sixteenth-century Spain was the same Arabism. How individuals used Arabic in different settings in Spain varied widely, as did the competencies they needed, and the consequences they faced. For example, in addition to its use as the daily vernacular of hundreds of thousands of Spanish subjects[4] and its status as a popular cultural marker employed for songs and spells,[5] Arabic was also an economic vernacular upon which both the Spanish treasury and many individuals depended.

In 1567, one of the most famous *morisco* advocates of linguistic relativism, Francisco Nuñez Muley, informed Philip II and his councillors that most tax, commercial, and property transactions in Granada were negotiated and recorded in Arabic.[6] Since Arabic was still very much the local language of record for the Granada economy, depriving Granada's *moriscos* of their right to use Arabic (as Philip had just done with his infamous *Pragmática* at the end of 1566), would have also deprived its inhabitants — and their ruler, Nuñez Muley was quick to insist — of the economic engine of the region.

Of course, Arabic was (and is) also a sacred language to Muslims and an important scriptural language for Christians too.[7] For this reason, studies of Arabic in early modern Spain — if considered within the historiographical frameworks of *morisco* or Catholic Reformation Spain — have, unsurprisingly, focused on the religious motivations behind learning or teaching Arabic, and on religious texts that were used for instructional, polemical, or evangelical purposes. Making the study of Arabic in non-sacred contexts even more complex is the fact that, in many ways, the religiosity of the age cannot be disassociated from the language used, in whatever genre of text. This is as true of references to Jesus or 'Our Saviour' in the matter-of-fact process of writing a date or the addressee in Spanish political correspondence, as of the Islamic features condemned by the Inquisition in Arabic marriage contracts between Valencian *moriscos*. However, there is a useful distinction to be made by contemporary scholars between how religious culture was communicated through different languages, and the specific genres of religious texts in explicitly (sacred) registers.

Indeed, the concept of register is fundamental for understanding how Arabic was studied, used, and proscribed in sixteenth-century Spain. We may take it for granted that all languages have different registers, although scholars rarely specify which register they are referring to when making claims about language

---

4 Pallarés, '"Algarabía" y "algemía"'; Vincent, 'Reflexión documentada sobre el uso del árabe'; Harvey, *Muslims in Spain*.

5 López-Morillas, 'Language and Identity in Late Spanish Islam', pp. 195–96.

6 Garrad, 'The Original Memorial of Don Francisco Nuñez Muley'.

7 García-Arenal and Rodríguez Mediano, 'Sacred History, Sacred Languages', pp. 133–36.

and history. To be sure, scholars of Arabic in Spain regularly account for the diglossic nature of a language between *fuṣḥā* and *ʿāmmiyya*.[8] However, beyond diglossia lies an infinite series of social-cultural registers, which also marked social boundaries which were not often crossed. Then, as now, most speakers of any language confronted a familiar set of registers or varieties that they had been equipped to deal with through their education and other experiences. Scholars of Spanish Arabism must account for these linguistic varieties in their analysis, and make clear which registers their sources and subjects are engaging with. And of course, in early modern Spain itself there was also a group of people who were confronted with a variety of Arabic registers, and whose professional qualifications rested moreover on being able to distinguish between those registers and react appropriately. These were Spain's Arabic translators, a heterogeneous category of experts called on by the administrative institutions of the Spanish monarchy in widely different contexts.

For professional experts — like Spain's Arabic translators — changes in register would have been far from anodyne. For example, when called upon to distinguish a Qur'ān from a copy of Avicenna's medical encyclopaedia, *al-Qānūn fi-l-ṭibb*, Inquisition translators had the relatively easy task of concluding a passive evaluation which would become part of a collection of evidence to be used in trials and sentencing, after which time there would be little need to refer to such translation. Indeed, they used the term *calificar* (assess) to characterize their reading of and reporting upon the potential usefulness or danger of Arabic texts, rather than *traducir* (translate). Likewise, military translators working on the far side of Spain's Mediterranean frontier conducted negotiations, inquests, and agreements while on campaign that were largely personal and oral, and most likely in a variety of *ʿāmmiyyāt* (dialects) or a specialized military register between classical and vernacular uses, approximating what scholars would now refer to as a 'Middle Arabic'.[9] Although the effects of this Arabic-Spanish translation were immediately significant for the provisioning and security of their garrisons, the translations themselves were ephemeral and rarely recorded. On the other hand, the *romanceadores* of Spain's appellate courts were called upon to make written translations of formal Islamic legal documents, whose legitimacy was intended to have a legal standing which could be (and often was) debated in courts of law for generations afterwards.

Thus, Spain's Arabic experts were highly specialized, and selected because their Arabic knowledge and experience had equipped them for particular tasks in specific registers. However, such specialization did not obviate the potential, or the need, for continuing Arabic education and for expanding any given translator's duties. This chapter examines just such a case, discussing the materials and strategies used by one of sixteenth-century Spain's most

8 García-Arenal and Rodríguez Mediano, 'Sacred History, Sacred Languages', pp. 135–36.

9 Lentin, 'Middle Arabic'.

famous Arabic experts, the *morisco* Alonso del Castillo. By focusing on the marginal annotations compiled by Castillo in one of his workbooks (Madrid, BNE, MS 7453) — which have never been studied — I will discuss what such working materials reveal about how an Arabic expert in sixteenth-century Spain approached the task of expanding his linguistic competencies, both out of a sense of genuine intellectual curiosity, and for a chance at professional advancement. Looking in detail at these unpublished but collated working materials tells us how Castillo's professional strategies changed over time, as his knowledge of a new register — Moroccan chancellery Arabic (*inshāʾ*) — improved, in tandem with his rising profile within the Spanish administration.

Studying the relatively well-known case of Castillo 'from the margins' shows how professional Arabists in Spain could distinguish themselves from others with different knowledge or experiences of Arabic by cultivating linguistic skill as an ongoing process. As Castillo himself noted in a letter to Gabriel de Zayas, one of Philip II's royal secretaries,

> entonces tendría por bien merecido el estudio de la gramática arábiga, pues tan bien se ha venido a emplearse en servicio de tan gran Rey y señor nuestro y en el de V[uestro] m[erce]d'.
>
> (thus I find it well worth studying Arabic grammar, because I have used it so much in the service of such a great King and in that of your lordship.)[10]

Effectively, this meant that one of the hallmarks of the Arabic expertise of someone like Castillo was, in fact, the limits of his knowledge, his desire to continue learning, and his ability to adapt those limits based on the needs of his patrons. This was achieved not only through the continued study of grammar. Indeed, recovering Castillo's marginal notes and the materials he interpolated among the translations also adds to our knowledge of what Arabic lexicographical materials were available in early modern Spain, and paints a new picture of: 1) where Arabic fits into the broader history of European lexicography, and 2) where Spain fits into the broader picture of Classical Arabic lexicography. Ultimately, Castillo embodied a parallel lexicographical tradition which combined a continuity of Classical materials with European vernacular trends, yielding an alternative narrative of Arabic and Arabism in early modern Spain.

The Continuing Education of an Arabic Translator in Habsburg Spain

Castillo's life has been studied in great detail by Dario Cabanelas, whose thorough biography — the result of a series of articles about Castillo published

10 Madrid, Biblioteca Nacional de España (hereafter BNE), MS 7453, fol. 59r, *apud.* Cabanelas, 'El morisco granadino Alonso del Castillo, intérprete de Felipe II', p. 35.

over two decades — remains the first port of call for scholars interested in any of Castillo's myriad activities.[11] Mercedes García-Arenal and Fernando Rodríguez Mediano have continued to advance our understanding of Castillo's biography, his family and patronage networks, and the social and political contexts (local and royal) in which he acted (or aspired to act).[12] He was born in around 1520 in Granada, to a *morisco* family with some professional tradition in healthcare and medicine. He was probably the son of Hernando el Castillo, 'al-Achal' (*al-Akhāl*) or 'el boticario' (the apothecary). Later testimony from the *morisco*-turned-Moroccan-translator Aḥmad ibn Qāsim al-Ḥajarī refers to one of Granada's late sixteenth-century translators as *al-Ukhāyāl*, or the diminutive of *al-Akhāl*.[13] Castillo went on to study medicine, and sometime around 1540 he received at least an introductory degree (*licenciatura*) to practice.

A second-generation *morisco* in Granada, Castillo's native language was the Granadan dialect, though he also came to speak Castilian fluently, learned some Latin and even a few Greek words, most likely during his university studies. His classical Arabic education, as he reported during later inquiries surrounding the apocryphal *plomos*, did not take place among Granada's remaining (crypto)-Islamic scholars, but with the Flemish humanist Nicholas Clenardus, who was in Granada during the late 1530s.[14] By the time the latter met Castillo, the highly-motivated Clenardus would have been on his journey of Arabic learning for over two decades, with access to a range of classical Arabic grammatical works and other texts common for learning Arabic among sixteenth-century European Christians, like an Arabic version of the Gospels.[15] We have no record of what materials Castillo studied with Clenardus, or on his own, but he was probably exposed to some of the classical works of Arabic grammar circulating in the region, like the *Alfiyya*, the *Ajurrūmiyya*, or the *Kāfiyya*. These works formed the basis for Arabic study across the Peninsula, whether by *moriscos* or foreign scholars.[16]

Arabic studies in Spain were thus underpinned by a 'store of grammatical knowledge' derived directly or through compendiums and commentaries from the Classical Arabic tradition.[17] However, in sixteenth-century Spain there was no regular means to study Arabic in the universities or otherwise, which is what had taken Clenardus to Granada in the first place, searching for Arabic expertise

11 The bibliography is compiled in Juan Martinez Ruíz' preface to the 1991 reedition of Cabanelas' monographic study, *El morisco granadino Alonso del Castillo*, pp. 20–27.

12 Especially in García-Arenal and Rodríguez Mediano, *Un Oriente español*, and the second edition in English, García-Arenal and Rodríguez Mediano, *The Orient in Spain*.

13 García-Arenal and Rodríguez Mediano, *The Orient in Spain*, p. 103.

14 García-Arenal and Rodríguez Mediano, *The Orient in Spain*, p. 105.

15 Muñoz, 'The Teaching and Learning of Arabic in Salamanca', pp. 178–81 Chorão Lavajo, 'Itinéraire Islamo-arabisant de Clénard', pp. 293–96.

16 García-Arenal and Rodríguez Mediano, 'Sacred History, Sacred Languages'; García-Arenal and Rodríguez Mediano, *Un Oriente español*; Zánon Bayón, 'Los estudios de lengua árabe'; Fournel-Guérin, 'Le livre et la civilisation écrite'.

17 García-Arenal and Rodríguez Mediano, 'Sacred History, Sacred Languages', p. 161.

after trying at several universities and humanist libraries in Salamanca, Evora, and Seville for nearly a decade in the 1530s. By 1543, the University of Salamanca had developed a piecemeal Arabic curriculum in which students were taught from the *Ajurrūmiyya* by the Hebraist Martin de Cantalapiedra.[18] Arabic study would not become established in Spanish universities until the arrival of Castillo's successor, Diego de Urrea, whom Philip II installed in the Arabic chair at the University of Alcalá de Henares in 1593.[19] Even then, Urrea's tenure ended with his retirement to Naples in 1611, and both the demand for and the supply of formal Arabic instruction in Spanish universities remained low in the sixteenth and seventeenth centuries, leaving interested scholars to seek out private tutors.

Meanwhile, some *moriscos* continued to organize Arabic instruction throughout the sixteenth century. Prospective students among the non-Arabophone *morisco* populations of Castile or Aragón might have wished to access classical texts to maintain the cultural and religious (and legal) traditions of their communities. For this market, networks of Aragonese and Granadan *alfaquíes* and other linguistic experts attracted itinerant students, like the Mancebo de Arévalo, or undertook didactic tours themselves to help ensure the transmission of specific expertise in *morisco* communities across the Peninsula.[20] In contrast, for native speakers of the various *'āmmiyyāt* spoken in the Peninsula, their motives to learn *fuṣḥā* would have been, in the majority, aligned with this first group. Meanwhile, other prospective students who wished to read the Qur'ān or otherwise access Islamic texts may have been Christian clergy invested in polemic or evangelical encounters with Muslims — in Spain or overseas. For these students, sympathetic or, better yet, converted *alfaquíes* were important teachers and linguistic informants.[21] Outside the *alfaquí* networks and the universities, religious orders like the Jesuits promoted the study of Arabic by founding dedicated schools, including the Collegio Romano, where several Spanish Jesuits would study *fuṣḥā* in the 1560s before returning to Spain to work in the *morisco* mission.[22]

---

18 Teaching Arabic in Salamanca (and other major European universities) had been mandated in 1311 at the Council of Vienne, though the first evidence of this order being in effect in Salamanca is only in 1411. Actual records of Arabic teaching and teachers do not appear until the 1540s and, even then, are quite elusive. Muñoz, 'The Teaching and Learning of Arabic in Salamanca', pp. 175–78. For Cantalapiedra's career, see López-Baralt and Iverson, *'A zaga de tu huella'*, pp. 98–107.

19 On the Arabic chair at Alcalá de Henares, and Urrea's career more generally, see García-Arenal and Rodríguez Mediano, *The Orient in Spain*, pp. 227–28.

20 Narváez-Córdova, 'Writing without Borders'; Harvey, *Muslims in Spain*.

21 The most famous examples may be Iça de Gebir (fl. 1450) and Juan Andrés (fl. 1515). See the studies by Wiegers, *Islamic literature in Spanish and Aljamiado*, pp. 69–150, and Szpiech, *Conversion and Narrative*, pp. 32–41. An important example only recently brought to light is Juan Gabriel of Teruel, a *morisco* convert to Catholicism who was previously the *alfaquí* Ali Alayzar. See Starczewska and García-Arenal, '"The Law of Abraham the Catholic"'.

22 On the Jesuit Arabic programs of the 1560s, see Gilbert, *The Politics of Language in the Western Mediterranean*, pp. 308–13, as well as El Alaoui, 'El jesuita Ignacio de las Casas y la defensa de la lengua árabe', p. 19. On the Collegio Romano and the context of Arabic study in Rome, see Girard, 'Teaching and Learning Arabic in Early Modern Rome'.

Eventually, in the seventeenth century, Arabic scholarship in Spain would officially centralize around the collections of the Escorial library, whose first catalogue Castillo had been charged to make in 1577, and scholars working in or in contact with Granada about the *plomos*.[23] Nonetheless, when Castillo was visiting the Escorial and corresponding with other scholars in connection with the library and the court in the 1570s and 1580s, there were no known competent Arabists other than himself to study with. One important characteristic of this time which informed the ambivalence around Arabic study in Spain at this time was, of course, the legal proscription against possessing Arabic texts. As a *morisco* from Granada, Castillo was among the Spanish population for whom this proscription was actually enforced.

Thus, when Clenardus and Castillo met in 1530s Granada, which had experienced recent and repeated royal edicts prohibiting the use of Arabic or the possession of Arabic texts, continuing Arabic study would have been a risky proposition for a young *morisco* like Castillo.[24] So what motive would have compelled Castillo to improve his classical Arabic? Knowledge of classical Arabic would have been important for any Muslim scholar, but Castillo was a baptized Christian, who had received an education rooted in Christian humanism at the University of Granada. As a medical professional, and as a translator originally specializing in legal documents, the classical register would have been of great interest to Castillo, even if he lacked access to a complete classical education. Could he have also been interested in more traditional classical Islamic discourses? Indeed, his personal religious beliefs remain a subject of debate among scholars, though he received written support from the highest ecclesiastical quarters upon his death, testifying to his true Christian belief and his right to be buried in a Christian cemetery.[25]

Castillo was a lifelong student, and his studies were strategically linked to his professional advancement as a translator. In the first decades of *morisco* Granada, a veritable industry of translation had exploded. We may be confident that a significant amount of *ad hoc* translation and different degrees of bilingualism in everyday interactions occurred during these decades. However, a specific professional niche for Arabic experts was also created and institutionalized within the overlapping institutions of power and government in Granada, the town council (*cabildo*), the military governor (*capitanía mayor*), the church and the king or his representatives. Castillo must have seen an opportunity not only for additional income, but a route to move closer to the corridors of power — an exercise he would spend a good part of his life trying to achieve.

23 Rodríguez Mediano, 'Fragmentos de Orientalismo español del s. xvii'; Hershenzon, 'Traveling Libraries'.

24 The older generations would have been protected to some degree from the authorities by their ability to claim that whatever Arabic they knew dated from before the conquest and the legal proscriptions against the language.

25 García Pedraza, *Actitudes ante la muerte en la Granada del siglo XVI*, v. ii. 1010–11.

In 1556, the longstanding municipal interpreter (*romançeador*), Juan Rodríguez, died, creating a need for new Arabic-speaking personnel.[26]

For over a decade, between 1556 and 1568, Castillo worked regularly for the town council and the *Real Chancillería* (Royal Appellate Court). In addition to translating Arabic legal documents, in 1564 he was commissioned by the council to conduct a kind of archaeological survey of Granada's remaining Islamic monuments — like the palaces of the Alhambra — and to make translations of their Arabic inscriptions. Although Cabanelas corrects Castillo's not-infrequent orthographic and other faults in the *morisco*'s copies of these transcriptions in Madrid, BNE, MS 7453, the latter's interest in these poems and his ability to copy and translate them also implies some working knowledge of Arabic poetry and poetics.[27] At the same time, he also undertook to translate inscriptions in private homes, like that of his friend Hernando el Chapiz.[28] Across these different institutions, Castillo was able to draw upon a breadth of experience with the Arabic language and its different registers, but he was also often confronted by the limits of his knowledge and the need to keep learning. Ultimately, these many experiences allowed him to participate — to what degree it has so far been impossible to prove definitively — in the creation and translation of the apocryphal Arabic, Latin, and Spanish prophetic and pseudo-doctrinal texts that made up what is now referred to as the *Libros plomos* (Lead Books).[29] The translations and evaluations of these texts, which began to be discovered by workmen, amateur archaeologists, and other treasure hunters in 1588, with the most important cache coming to light in 1595, would dominate the last decades of Castillo's life in Granada.

Before Castillo played the starring role in the end-of-century translation movement around the *plomos*, however, he had already spent decades as a professional Arabic expert in the service of various institutions across the Spanish monarchy's overlapping administrative jurisdictions (municipal, ecclesiastical, royal). Through it all he displayed a consistent attention to language and specifically to cultivating (and advertising) his own linguistic expertise. While translating legal documents for lawsuits to be heard in the *Real Chancillería*, he did not hesitate to correct the work of previous

---

26 Archivo Histórico Municipal de Granada, Libro V, fol. 270^r. Alonso del Castillo, 'Sumario e recopilación de todo lo romançado', p. 25.

27 See the bibliography in Cabanelas, 'El morisco granadino Alonso del Castillo, intérprete de Felipe II', pp. 20–27.

28 On their friendship, see García-Arenal and Rodríguez Mediano, *The Orient in Spain*, pp. 98–102.

29 For an overview of Castillo's role in this affair, the context around it, and an up-to-date bibliography, see García-Arenal and Rodríguez Mediano, *The Orient in Spain*, pp. 95–120. Gerard Wiegers and Pieter Sjoerd van Koningsveld have continued to advance new research about the *plomos* including the publication of translations, critical editions, and photoreproductions; see for example van Koningsveld, 'Le parchemin et les livres de plomb de Grenade', pp. 171–96, Wiegers, 'El contenido de los textos árabes de los plomos', pp. 197–213, and van Koningsveld and Wiegers, 'Apéndice documental', pp. 215–58. See also by both authors, *El pergamino y los libros plúmbeos del Sacromonte*.

Arabic translators, especially their lexical precision.[30] While describing his experiences translating for Juan de Austria's armies in the War of the Alpujarras (1568–1570), he paid particular attention to the details of writing in Arabic with the appropriate diacritic marks, showing the correct declension and conjugation (*ḥarakāt* and *tashkīl*), and praising it for being 'bien xuclada' (a Spanish loan word from Ar. *shakala*, 'to provide with vowel points', but adopting Spanish morphology, e.g., the participle form *xuclada* rather than *mashkūl* or *mushakkal*).[31] In his position with the Granada Inquisition he displayed his erudition by identifying some works of science and political theory to the king and his secretaries as being appropriate for the royal collections at the Escorial, which was then being constructed.[32]

In the 1580s, however, with over two decades of professional experience and recognition already to his credit, Castillo was faced with a new challenge that would take him to the limits of his Arabic knowledge. In 1579 he was called upon to translate the correspondence between the new sultan of Morocco, Aḥmad al-Manṣūr, and Philip II. Tasked to render the language of the Moroccan chancellery into the language of the Spanish *consejos*, he encountered texts unlike any he had seen before. The chancellery register used by the Moroccan *makhzan* was a sophisticated political and religious discourse which served to represent the sovereignty of its ruler abroad. Although Castillo had translated letters during the Alpujarras war, those had been rhetorically more straightforward than the elaborate epistles sent by al-Manṣūr.[33] Translating the Moroccan texts was a much higher-stakes endeavour — both personally for Castillo's own ambitions, and for the fraught international relations they would come to underpin — than any Castillo had previously undertaken.

A Marginal Correspondence

Between 1579 and 1587, Castillo recorded translating 51 Arabic letters from Moroccan and Ottoman agents in the Mediterranean which were sent to Spanish officials or fell into the hands of Spanish agents. The Moroccan (and Ottoman) letters as recorded in Castillo's workbook have almost all been translated and edited, either by Dario Cabanelas in one of his numerous articles about Spanish-Moroccan relations under Philip II written between 1956 and 1974, or in the 2002 document collection *Cartas marruecas*.[34] Only three of the fifty-one letters have never been edited. In all cases, however, Castillo's

30 Merino, 'La traducción de cartas árabes en un pleito granadino del siglo XVI', esp. pp. 511–12.

31 Alonso del Castillo, 'Sumario e recopilación de todo lo romançado', p. 25.

32 Cabanelas, 'El morisco granadino Alonso del Castillo', pp. 136–37.

33 In his Alpujarras translations, Castillo translates almost every time the *duʿāʾ* as 'Loado sea Dios!' and there are no complex elegiac greetings between the *morisco* leaders as he would find in the Moroccan and Ottoman correspondence he was later tasked to translate.

34 García-Arenal, Rodríguez Mediano, and El Hour, *Cartas marruecas*, docs. 1–8.

copious marginal annotations have gone unstudied. It is these annotations and the other materials included in the workbook that give a special insight into his lifelong Arabic education and especially how he used translations and lexical study to enhance his expertise and professional qualifications.

As scholars have pointed out, Castillo never achieved a total fluency of style in his Moroccan translations into Spanish.[35] Nonetheless, the differences between his first translation — when he had little experience of the Moroccan chancellery register — and his later translations show evidence of Castillo's learning process, both in Arabic and in the rhetorical expectations of the Spanish court. He translated the first letter from the new Moroccan Sultan Aḥmād al-Manṣūr to Philip II in March 1579. In this instance, exceptionally, Castillo managed to make a copy of both the original Arabic letter and his Castilian translation in the workbook.[36] In the Arabic version, he filled the margins with notes about vocabulary, like one of the epithets ascribed to Philip with which he was not familiar, *al-manāqib*. This term, which can mean a mountain trail or pass, in al-Manṣūr's letter referred generally to Philip's virtues. Castillo noted this bivalency in the margin next to the text where the word appeared, having indicated the term in the text with an overline:

المَنْقَبُ و النَقْب [هو] الطريقةُ في الجَبَلِ
و مَناقِبُ الانْسانِ [هو] اوْصافَهُ و ظُروُفَهُ التي يَسْماَىَ عَليهَا و يعيشُ فبها

> (*al-manqab* and *al-naqb* [is] a path in the mountains
> while the *manāqib* of a person [are] his characteristics and circumstances,
> by which he is called and by which he lives)[37]

It is also interesting how he then translated such new vocabulary. The fuller phrase of this particular epithet was 'to the great king […] possessor (*ḍī*) of […] exquisite virtues (*al- manāqib al-ḥisān*) […]'. In the Spanish version, Castillo did not translate this as *virtudes hermosas* or another kind of generality regarding virtue, but instead listed three specific virtues that were not literally present in the Arabic original: 'prosapia e generosidad y espléndiods hechos' (lineage and generosity and splendid deeds). Thus, as Castillo learned the new meaning of a familiar word, he also used a creative translation strategy to ensure that this meaning would be received by the Spanish king in the way it had been intended.

Another revealing example of how Castillo used marginal notes to improve upon his knowledge of certain words comes from the same letter.

35 The anonymous author of the 1852 preface to Castillo's 1574 *Recopilación* is withering on this point, see Alonso del Castillo, 'Sumario e recopilación de todo lo romançado', p. 7.

36 Castillo was closely supervised while translating the sensitive Moroccan correspondence, working in tandem with his patron Pedro de Deza in Granada in order to produce acceptable Castilian documents to be sent back to Madrid. Of the 51 translations, only a handful (5) are extant in both Arabic and Castilian versions.

37 Madrid, BNE, MS 7453, fol. 271^{v}. Castillo's *ḥarakāt*.

When discussing the liberation of Philip's commander Juan de Silva, Castillo transcribed:

مككنا خديمكم جوان ذي شِلْبا

when he should have written:

فككنا خديمكم جوان ذي شِلْبا

(we liberated your servant Juan de Silva)

Indeed, Castillo wrote in the margin below, correcting either his own transcription or the original:

مككنا ... اظنُّهُ فكَكْنا بالفاء

(*makakna*. I think this is *fakakna*, with 'fa')[38]

Here Castillo corrects the text, and explains both his corrections and the significance of them (in this case, to craft a phrase that makes sense).

From 1580–1581 Castillo copied only the Castilian summaries of the letters he translated, though he included notes about 'los más obscuros e notables terminos' (the most obscure and remarkable terms) from those letters recorded in lists after the summaries.[39] As will be discussed below, the latest letter for which Castillo made copies of both the Arabic original and his Spanish translation was processed by Castillo in December 1583. From 1584 onwards, he copied only the Arabic originals, although he continued his copious marginal notations. One simple example from an early letter, translated on 2 February 1580, shows how Castillo perhaps intended to arrange his marginal notes and lexical references (though he would not remain so organized in subsequent letters).

This letter, for which he gave an extensive summary that is almost a paraphrase rather than a summary or translation, Castillo characterized as 'una capitulación que en arabiga quiere decir Jarida' (a capitulation, which means Jarīda in Arabic) and in the margin he wrote 'جريدة'.[40] Then, after concluding a series of eleven Spanish summaries, he made a list of 'notable terms', the first of which was 'جريدة الفصول'.[41] To explain this term he made notes in Arabic, Spanish, and Latin:

الجريدة (هي) البرنمج الذي تتجرّد فيها المعلن عن سيارها. من جرد يجرد

(*al-jarīda* (is) the index through which information is divulged as it (the index) progresses. From *jarada yajridu* [to make an inventory])

---

38 Madrid, BNE, MS 7453, fol. 270^{v}.

39 Madrid, BNE, MS 7453, fols 7^{r}–8^{r}, 10^{v}, 14^{r-v}.

40 Madrid, BNE, MS 7453, fol. 5^{r}.

41 Madrid, BNE, MS 7453, fol. 7^{r}.

> Es copia o capitulación o memorial (it is a copy or a list of articles or a report)
>
> Copia en romance (in Romance language, *copia*)
>
> Latine proprie Ellencus (in Latin, *elenchus* [catalogue, index])[42]

Though a basic term, this entry is the first and the most thorough of all of Castillo's annotations of the Moroccan correspondence, in that it gives the Arabic definition, refers to the Arabic root and verb form (as would have been customary in classical grammar), and provides a definition in both Spanish and Latin. Never again, it seems, would Castillo have sufficient time to fill out his lexical list so comprehensively. Nonetheless, this entry gives an insight into how Castillo imagined composing and using such a lexical tool.

Aside from his marginal notes and comments, some of the ways in which Castillo's understanding of chancellery Arabic improved were to do with gaining familiarity with the standard components of a diplomatic letter and the turns of phrase which characterized the Moroccan chancellery style.[43] For example, over time Castillo grew more adept at translating the repetitive formulae of greetings and mutual praise that comprised the first third of most letters between Muslim rulers. For instance, in 1579 he translated the elegiac preface of al-Manṣūr's response to Philip *de verbo ad verbum* — resulting in a singularly tortured and difficult-to-understand Castilian full of Islamic epithets and references which would have meant little to Philip's secretarial staff or to the king himself. This included a formulaic component of almost all the letters issued from the Moroccan chancellery, a version of the following prayer to the prophet Muḥammad:

> الصلاة والسلام على نبينا محمد وكافّة الرسل والانبياء، بدور الهدى وصفوة الخلق في الأرض وفي السماء.
>
> (Prayer and peace to our prophet Muhammad — and to the entirety of the messengers and prophets — the pole of righteousness and the best of creation on earth and in heaven) [44]

42 Madrid, BNE, MS 7453, fol. 7^r^. 'Elenchus' is likely a Latin loanword derived from Gr. *elenkhos* (argument, proof) which has been imported into Castilian as *elenco* (catalogue, index). Castillo defined it elsewhere in the workbook in another marginal note as 'la tabla o argumento del libro', s. fol. 2^r^. On the *Elenchus* or *Series titulorum* used in early modern reference books from the same period, see Blair, *Too Much to Know*, p. 337.

43 On al-Manṣūr's chancellery, its staff, and practices, see Mouline, *Le califat imaginaire d'Ahmad al-Mansur*, pp. 212–33, esp. p. 229.

44 Madrid, BNE, MS 7453, fol. 271^r^.

Castillo translated this quite literally but with some errors as:

> salud y santificaçión con nuestro Profeta Mahamad e con todos sus Mensageros e Profetas, luzeros[45] de la honestidad e limpieza de los hombres que son en la tierra y en el cielo.[46]
>
> > (peace and sanctity with our prophet Muḥammad along with all his messengers and prophets, illuminated with the honesty and purity of those men who are on earth and in heaven)

What Philip and his secretaries would have made of these Castilian renditions of praise for Muḥammad — at the very same moment that the Inquisition was confiscating items with similar phrases and punishing those found with such materials — is not known. Castillo's own reflection on this question may be what prompted him to adjust his translation strategies for future letters.

Indeed, by 1583 Castillo had developed a far more efficient and synthetic style for his Castilian translations. In a translation made on 16 November 1583, he went so far as to elide whole sentences and paragraphs from the original letter, recombining phrases and ideas into a more efficient Castilian introduction.[47] Notably, he eschewed conventional Islamic epithets like 'أيد الله امره وأعزّ بتأييده نصره' (may God support his rule and glorify the endorsement of his victory) that were given in the Arabic original, providing no Spanish translation for these passages.[48] In 1579, Castillo had translated a comparable epithet at the same point in the letter (after introducing al-Manṣūr as the sender), 'احسن الله اليه وافاض نعمه ظاهرة وباطنة عليه', (may God favour him [al-Manṣūr] and confer blessings upon him inside and out) in his literal translation of the elegiac preface as '¡Dios le haga bien y le cumpla de sus gracias notorias y ocultas!' (May God provide him rewards and grant his graces both public and private) while, in 1583, he entirely eliminated phrases of this type, which were after all reminders of the obligations of al-Manṣūr's own secretarial staff.[49] Both approaches were a far cry from Castillo's earlier strategy of translating all epithets as *¡loado sea dios*! (Praise be to God), as he had in his work in the Alpujarras.

Nonetheless, despite having converted to a *sensum ad sensum* approach, Castillo maintained essential diplomatic forms like claims of sovereignty and legitimacy, essentially rewriting the Arabic original into a format that was

45 Did he read 'nūr' for 'dawr'?

46 Madrid, BNE, MS 7453, fol. 2ᵛ.

47 The first half folio of Castillo's Spanish translation is a synthesis of nearly whole first folio side of the Arabic original, which is only one folio recto-verso in length, while the Spanish version (in much bigger letters) is nearly five folio sides long. Castillo took concrete information, such as the letter's origin in Marrakesh — information that was otherwise included in several lines of the *du'ā'* — and transposed it into the greeting and address. Madrid, BNE, MS 7453, fol. 45ʳ.

48 Madrid, BNE, MS 7453, fol. 45ʳ.

49 Madrid, BNE, MS 7453, fol. 1ᵛ for the Spanish translation and fol. 271ᵛ for the Arabic original of this phrase.

more acceptable and easily digestible for the Spanish chancellery. In the 1583 example, he did not ignore all of the copious elegiac titles which al-Manṣūr ascribed to himself and to Philip. These included, for al-Manṣūr, titles like *khalīfa*, which Castillo rendered simply as a loanword into Spanish rather than translating it: *el halifa excelente*.[50] Notably, in this case Castillo also added immediately after the phrase in Castilian *el Caudillo el Mahdi* where al-Manṣūr's scribe had written the conventional prayer *al-qā'im bi-amr Allāh* (he who is empowered [to carry out] the orders of Allah). In this case, Castillo edited — whether intentionally or intuitively — part of the Moroccan chancellery rhetoric which was intended to discursively situate al-Manṣūr in a position of superiority to his correspondent, a frequent tactic in his correspondence with European rulers.[51] However, at the same time he used Castilian terms for military and North African leaders with which Philip's councillors would have been familiar.[52] Philip himself was fashioning himself in an analogous way as 'King of Jerusalem'[53] and would have recognized the significance of this title as Castillo rendered it. Later in the same letter, Castillo translated the following address:

> الى المثاب التي لها في الامام المسيحية صيب الذكر والاشهار، وتنويه المقدار، مثابة السلطان الماجد الافخم، الاسنى، الاسمى، الازكى، الارضى، الاحظى، الاثير، الخطير، السلطان ذون فلبي بن السلاطين الكبر، الذين لهم تمام الصيب و تعظيم الاشتهار
>
> (to the court where the head (*imām*) of Christianity is the greatest in renown and extensive tributes, the court of the glorious and splendid Sulṭān — the most resplendent, the most exalted, the most intelligent, the most pleasant, the most favoured, the beloved, the important (*al-khaṭīr*)[54] — the Sulṭān Don Philip, son of the great sultans who had complete purity and immense fame)

as:

> la Grande e más Sublime Magestad e Tribunal de la ley del Mexías, que es el rey Don Phelipe, hijo de los muy grandes e aventajados reyes en ser e nombre e poderio

---

50 Madrid, BNE, MS 7453, fol. 45r. Significantly, al-Manṣūr had not used this title in 1579, as he consolidated his power in Morocco and waited to see what measure of threat or ally the Ottoman sultan — who also claimed the supposedly unique title of *khalīfa*. See Cory, *Reviving the Islamic Caliphate in Early Modern Morocco*, pp. 23–24.

51 Mouline, *Le califat imaginaire d'Ahmad al-Mansur*, p. 229.

52 On the figure of the *mahdī* in Morocco and al-Manṣūr's claims as *mahdī* as well as *khalīfa* (though he never used the former term in his correspondence), see García-Arenal, *Messianism and Puritanical Reform*, pp. 269–95.

53 García-Arenal, *Messianism and Puritanical Reform*, p. 277.

54 There is some ambivalence in this choice of words, which also means 'dangerous'.

> (the great and most sublime Majesty and Tribunal for the law of the Messiah, who is King Philip, son of those very great kings, noble in being, name, and power)[55]

In his translation Castillo eliminated many of the repetitive epithets, though he still translated enough of the elegiac greeting to express al-Manṣūr's diplomatic politesse and the intended relative position of the two monarchs.

In this way Castillo, like other translators at the time, was also exercising a secretarial function in *composing* — rather than simply translating — the letters that were then read by the king and his councillors.[56] In Castillo's case, he was at least for a time aided in this portion of his translator's task by other royal officials, like his patron Pedro de Deza, with whom he would review his Castilian translations and make a 'clean' copy (*en limpio*) before sending it on to the court. Part of this process seems to have been removing invocations (*duʿāʾ*) that were not perceived as being essential to the letter's message.

In addition to learning how best to manage the rhetoric of diplomacy and alliance for his royal audience, Castillo was also confronted with the task of establishing Muslim and Christian date conventions in the correspondence. Despite his extensive experience, he was never wholly at home with making the chronological conversion between Muslim and Christian dates. In his previous work as a translator of Islamic legal documents for the *Real Chancillería de Granada*, Castillo had also been called on to translate Islamic dates for Christian institutions.[57] The exactitude of Castillo's chronological conversion for medieval property records, however, was far less pressing than giving a correct date for the (relatively) rapid Mediterranean exchanges which conditioned the rhythm of war and peace across the sea.

He seems to have realized the importance of converting the dates to the sequence of correspondence in international diplomacy. Indeed, the problem of chronological conversion was becoming a topic of vital interest to Spanish scholars by the end of the sixteenth century.[58] As part of Castillo's self-preparation to learn new skills as a royal (that is, chancellery) translator, he also set out to study this question of chronology, as evidenced by several folios of MS 7453 that are devoted to discussing the Islamic calendar.[59] Although he never wholly mastered the conversion (which may have been

---

55 Madrid, BNE, MS 7453, fols 48^{r-v}.

56 Al-Manṣūr's own Spanish translation staff were quite adept at this kind of 'translation diplomacy' in the Spanish versions of al-Manṣūr's Arabic correspondence that they prepared for Spanish and English courts. See Gilbert, 'The Circulation of Foreign News and the Construction of Imperial Ideals'.

57 For example, in the 1558–1559 *pleito* over water rights studied by Manuel Espinar Moreno, Castillo was hired to translate some Nasrid parchments recording the sale of related properties, and for which Castillo miscalculated the chronological equivalence of various dates by several years. Espinar Moreno, 'Escrituras árabes inéditas del siglo XV', pp. 35–41.

58 Rodríguez Mediano, 'Fragmentos de Orientalismo español del s. XVII'.

59 Madrid, BNE, MS 7453, fols 3^{r} and 12^{v} in the unfoliated pages beginning the left-to right reading.

further complicated by conversion to the Gregorian calendar reform after 1582), he began to include a chronological conversion as part of his summaries and translations from 1581 on.[60] Nonetheless, these conversions remained approximate at best and, since Castillo sometimes even dated letters as being sent *after* he had translated them, it would be extremely interesting to know how this conflicting information was received and processed in the *consejos*.[61]

Aside from these formulaic issues, Castillo was also faced with a barrage of unfamiliar vocabulary. Throughout his workbook, he filled the margins of his copies of both Arabic and Castilian versions of each letter with relevant lexical notes. For certain letters, he also recorded his impressions about the quality of their style and their value to his continuing Arabic education, as in 1586 when he prefaced an Arabic copy with the note that he had copied it 'para annotaciones de sus terminos porque está muy sentida de terminos muy elegantes' (in order to annotate its terms because it is full of very elegant terms).[62] In this case, the letter was a significant missive in which al-Manṣūr went so far as proposing the surprising idea that Philip II's agent, Diego Marin ('nephew' of the original Diego Marin, Castillo's friend and Philip's *de facto* ambassador), should become the official intermediary of *both* rulers. The terms which seem to have most interested Castillo in this particular letter was the administrative term *durūba wa-mumārisa,* in reference to Marin's qualities in the information service and the potential mutual benefit.[63]

Through his marginal notes and the materials he collected and bound together with the copies of his translation work, Castillo created a lexicographical reference at various levels. His copious marginal notes show how he processed the letters he was set to translate, looking for unfamiliar terms and doing his best to situate them in context. When a word was potentially unfamiliar even in translation, he provided an additional explanation. For example, in 1583 when translating another letter from al-Manṣūr, he described in some detail the toponym 'Larache' (of crucial interest in Spanish-Moroccan diplomacy), both in the summary preceding the translation, and in the translation itself, explaining in this letter that al-Manṣūr (the *rey xerife*)

60 Madrid, BNE, MS 7453, fol. 12v.

61 For example, he described translating a letter from al-Manṣūr on September 16, 1581, 'cuya fecha se puso por el dicho rey moro en Veynte e dos dias del mes de Sabhen [*šaʿabān*] del año de la hijra de 989 q[ue] corresponde con los años de nuestro salvador Jesu xpo de 1581 en veynte e dos dias del mes de septiembre'. Madrid, BNE, MS 7453 fol. 12v. In fact, this date corresponds with September 30, not September 22. Without the original Arabic letter, we cannot be sure where the error was introduced.

62 Madrid, BNE, MS 7453, fol. 68v.

63 In the margin he explained, for his later reference, 'الممارسة العلاجة و المدارة للشيء دروبة' (practical experience (*al-mumārasa al-ʿalāja*) and the administration of things (*mudāra lilshay'*) are (known as) *darūba* (sic for Cl.A. *durba*)). He then added, 'الدروبة (هي) العادة و الجراءة على الحرب و كل امر' (experience is the readiness for the measures of war or any affair). Castillo borrowed this latter definition directly from al-Jawharī's *Ṣiḥāḥ* (see below). Madrid, BNE, MS 7453, fol. 69r.

a de determinarse en poner en efeto la dadiva de la fuerça de Anarache (*sic*) al rey Don Phelipe n[uest]ro s[eñ]or la qual en arabiga dizen guidala'rays q[ue] quiere dezir rio de las parras

> (must decide to effect the transfer to our lord King Philip of the fort of Larache, which in Arabic means *wādī al-ʿarāʾish* which means the river of the trellised vines.)[64]

Why he thought Larache — the Moroccan port-fortress whose possession was one of the key topics of Philip and al-Manṣūr's correspondence and Spanish-Moroccan diplomacy in general — would have needed this particular toponymic explanation is unclear.[65] However, it might have been that, while Larache was a familiar term in the *Consejo de Guerra* or *Estado*, it was less so in Granada, so the explanation was interference from Castillo's own lexical exploration.

Did Castillo master the chancellery register sufficiently to translate Spanish letters into Arabic? He had already translated letters from Philip into Arabic to be circulated among the *morisco* rebels during the War of the Alpujarras. In that case, the extant Spanish version of the letter Castillo translated into Arabic has no encomiastic preface — if for no other reason than because it was from the Spanish king, asking one of his rebellious subjects to surrender.[66] In the rest of the Alpujarras correspondence he translated, Castillo was never faced with anything so elaborate as the missives coming from Morocco. Despite his preparation in learning how to understand and translate this chancellery register, Castillo was probably never called upon to compose any Moroccan correspondence. Philip never showed any interest in making a diplomatic gesture like producing an Arabic missive, and he was well aware that he could rely on the presence of many Spanish and Portuguese captives and renegades in Morocco who he knew were serving al-Manṣūr in scribal as well as other roles. Indeed, from the 1590s on, much (though not all) of the formal correspondence issued from the Moroccan chancellery began to arrive accompanied by a Spanish translation. As al-Manṣūr's chancellery staff were far more capable and experienced in multilingual diplomacy than Philip's, for some time the Moroccan translations even began to obviate the need for an Arabic translator of that kind at the Spanish court.[67] The translator who succeeded Castillo as royal appointee — the Neapolitan renegade Diego de Urrea — spent most of the 1590s after his arrival setting up an Arabic programme

---

64 Madrid, BNE, MS 7453, fols 48^{r}–49^{v}.

65 On the negotiations over Larache during the reigns of Philip II and Philip III, see García-Arenal, Rodríguez Mediano, and El Hour, *Cartas marruecas*, pp. 47–110; Cabanelas, 'El problema de Larache en tiempos de Felipe II'.

66 Alonso del Castillo, 'Sumario e recopilación de todo lo romançado', pp. 35–37.

67 Gilbert, 'The Circulation of Foreign News and the Construction of Imperial Ideals', pp. 42–46.

at the University of Alcalá de Henares and managing the Arabic collections at El Escorial, rather than taking over the Mediterranean correspondence.[68]

Nonetheless, although Castillo's career as royal translator ended in 1587 — superseded in any case by his involvement with the *plomos* — his workbook from this period could have served as a vital resource for future translators.[69] The absence of other dates in the manuscript makes it impossible to know how it was used after 1587 (or how long before 1579 he had begun to compile it), and in whose possession it ultimately settled. However, some clues may be found by approaching the manuscript through the history of its archive. What is now BNE MS 7453 was acquired by the Spanish National Library in the 1750s as part of the collection of the Conde de Miranda, scion of the Zúñiga family that had played such key roles in the intellectual and political history of sixteenth-century Spain.[70] In the Miranda collection — whose own history is still poorly understood — there were six other Arabic codices in addition to MS 7453. These codices include a Spanish Arabic *Vocabulario,* annotated examples of the *Kāfiyya* and the *Ajurrūmmiyya,* both printed in 1592 by the Medici Press in Rome, and two versions of Arabic gospels, and García-Arenal and Rodríguez Mediano have concluded that they were produced in the context of the Arabic translation and study around the *plomos*.[71] In the Guillén catalogue a 1518 manuscript copy of al-Ghazālī's *Forty Principles of Religion* (*Kitāb al-Arbaʿīn fī uṣūl al-dīn*) is also attributed to the Miranda collection, although this book is not mentioned in Andrés' edition of the eighteenth-century inventory. With the exception of the al-Ghazālī, all of these codices seem to date from the period just after Castillo's activity in Granada. This would give them a provenance very near to Castillo's workbook. Whether the Miranda Arabic codices had been acquired together by the Zúñiga family or a previous collector, or one at a time, has so far been impossible to prove. Nonetheless, the uniformity of their theme and geography is a strong indication that they may have been at one time designed for use together.

Thus, it is possible that, although Castillo's workbook began as a personal reference book and professional record, it may also have provided the seed for a larger lexicographical project that would have been intended for a wider audience. Although far from a final draft, some sections of the manuscript show signs of revision and collation, notably the table of contents and the Moroccan *cartulario*. Throughout, he describes in detail the processes of drafting and writing new clean copies of the translations he had copied or summarized to

68 On this figure, see the biography and bibliography in Floristán, 'Diego de Urrea'. Under Philip III, Urrea would once again become intimately involved in Moroccan affairs, including negotiations over Larache, but this does not seem to have taken place under Philip II.

69 Many early modern reference books — both in manuscript and in print, or annotated — were intended to serve as a repository of useful knowledge for future generations of the professional or family network. See Blair, *Too Much to Know*, pp. 227, 290.

70 Andrés, 'Los códices del Conde de Miranda en la Biblioteca Nacional', pp. 612–15.

71 García-Arenal and Rodríguez Mediano, 'Sacred History, Sacred Languages', pp. 160–61.

be sent to the court. His earlier *Recopilación*, similar in format as a collection of letters and meditations on language, was recopied (by Castillo) *en limpio* in order to be formally presented to his patron Pedro de Deza in 1574 (Vid. Castillo 1852). Thus, Castillo was no stranger to the idea of manuscript publication, or publication in general. His translations had already appeared in print in Luis del Mármol Carvajal's *Historia del rebelión y castigo de los moriscos del reyno de Granada* (Málaga, 1600). In addition, Castillo's associate and fellow translator, Miguel de Luna (and a candidate for the possible provenance of at least some of the other Miranda volumes)[72] was famously able to produce a work of 'Arabic translation' for mass consumption in his *Historia verdadera del rey don Rodrigo* (Granada, 1592),[73] a pseudo-translation of the history of the Muslim conquest of Spain which was re-edited at least six times in the seventeenth century and translated into English, French, and Italian.[74] Hence, Castillo was hardly working on language-learning materials in secret, or within closed networks of crypto-Islamic education, like the remaining *alfaquíes* who were among the Inquisition's top priority by the 1580s. He created his materials through conversations and exchange with others, and it is likely that other Spanish Arabists were aware of his work and its potential for future students of the language.

If there was a potentially didactic aspect to Castillo's marginal annotations and his workbook, beyond serving as a personal record and reference, in what tradition should his materials be situated? Castillo has never really been considered as more than an ancillary in early Spanish scholarly orientalism, despite his connections to scholars like Ambrosio de Morales, or the events which fuelled the development of that orientalism, like the discovery and translation of the *plomos*. Nor is he associated with an enduring Spanish tradition of Classical Arabic learning. However, Castillo's workbook is in fact evidence of both a continuing interest in, and reliance on, key works of Classical Arabic lexicography that have previously not been thought to be circulating in *morisco* Spain. At the same time, his compilation of thematic vocabularies — including lists of medical and legal terms, sample letters and forms of courtly address, humanist maxims, and even diplomatic ciphers, resembles the trend in nascent European vernacular lexicography for producing specialized manuals with a range of didactic and sample materials. Just as Castillo's workbook bridged both traditions, a new narrative can emerge from it.

---

72 The *Inventario de manuscritos*, vol. XII, of the Spanish National Library considers that the volume 'Parece proceder del morisco Miguel de Luna'. Guillén Robles, *Catálogo de los manuscritos árabes existentes en la Biblioteca Nacional de Madrid*, p. 332. García-Arenal and Rodríguez Mediano do not agree that there is evidence in this volume for this provenance, however; see García-Arenal and Rodríguez Mediano, *The Orient in Spain*, p. 340.

73 Miguel de Luna, *La verdadera historia del Rey Don Rodrigo*.

74 Miguel de Luna, *Historia verdadera del Rey Don Rodrigo*, ed. by Bernabé Pons, pp. xxxiv–xxxvii.

A Translator's Handbook: Classical Arabic Lexicography and European Vernacular Lexicography Meet

Insights into the Continuity of Classical Arabic Lexicography in Morisco Spain

Despite its evident value for the study of Arabic learning in Spain, Castillo's workbook has been ignored in the main narrative of Spanish Arabic lexicography, a genre that is in any age closely connected with didactic methods. According to that narrative, thirteenth-century Spanish missionary manuals in Arabic gave way to Pedro de Alcalá's post-conquest lexicon of 1505[75] — a work which is generally characterized as either a last gasp of the medieval missionary tradition or — because of its emphasis on the Andalusi vernacular rather than classical Arabic — a detour in what would become an early modern orientalism that focused on classical texts and tools.[76] After Alcalá, according to the conventional narrative, Spanish Arabic lexicography would have to wait until the end of the eighteenth century to be resuscitated by the *ilustrados* in the courtly circles of Charles III and his minister Campomanes.[77] A rare exception, the Franciscan priest Bernardino González's *Intérprete arábigo*, a manuscript Arabic-Spanish dictionary dating to *c.* 1700, holds that Arabic instruction in Spain was primarily the preserve of the clergy.[78]

This narrative ignores two centuries of Spanish interest in Arabic. However, in the last fifteen years the activities connected to the study of Arabic in Spain have been recovered thanks to the work of Mercedes García-Arenal and Fernando Rodríguez Mediano, among others.[79] This scholarship incorporates important evidence of continuing Arabic study in Spain throughout the sixteenth and seventeenth centuries, in particular around the affair of the Lead Books (*plomos*) which, after all, lasted for nearly a century and drew in almost every Arabic expert in Europe. Nonetheless, even before the *folie* of the *plomos* overtook Spanish humanism, Arabic study was flourishing in several pockets of Spain, articulated around other endeavours like those discussed above.

What kinds of classical Arabic lexicographical works were circulating in sixteenth-century Spain? Al-Andalus did not cultivate a significant autochthonous lexicographical tradition, but the primary works in the

75 Corriente, 'La lexicografía árabe en España', pp. 133–39.

76 See for example, the discussion in Jones, 'Learning Arabic in Renaissance Europe (1505–1624)', pp. 21–22 and passim.

77 Vidal Castro, 'Los diccionarios español-árabe', pp. 321–25; Monroe, *Islam and the Arabs in Spanish Scholarship*, pp. 5–45.

78 González and Lourido Díaz, *Intérprete arábico*.

79 These scholars have opened the field and provided precious leads and guidance for many others, including this author. A current bibliography of some of their many contributions can be found in García-Arenal and Rodríguez Mediano, 'Sacred History, Sacred Languages'.

classical tradition certainly circulated and were adapted as needed.[80] These Andalusi transmissions echoed throughout the *mudéjar* and the subsequent *morisco* period. For example, the *moriscos* of Aragon had access to a *mukhtaṣar* (summary) of the *Kitāb al-ʿAyn* made by a tenth-century Sevillian scholar.[81] By the sixteenth century, the most significant and widely circulating Arabic lexicographical works in the Islamic tradition across the Muslim world were the *Ṣiḥāḥ al-ʿarabiyyah* by the eleventh-century author (and nephew of al-Fārābī), Abū Naṣr Ismāʿīl ibn Ḥammād al-Fārābī al-Jawharī, and the *Qamūs al-Muḥīt* by the fourteenth-century author, Muḥammad ibn Yaʿqūb al-Fīrūzābādī. Although these works do not appear — insofar as may be verified in the admittedly incomplete extant evidence — in the initial orientalist collections in Spain like those of Clenardus, or the Spanish humanists Páez de Castro, Hurtado de Mendoza whose books — including Arabic volumes — ended up forming the core of the Escorial Arabic collections, they did circulate in sixteenth-century Spain among *moriscos* and professional translators. For example, the Spanish National Library's copy of *Ṣiḥāh al-ʿarabiyya* was copied in Aragon in 1505.[82] At the end of the sixteenth century, there was a two-volume copy of the *Ṣiḥāh* in Granada available for the use of the *plomos* translators.[83] As a novelty at the turn of the seventeenth century, Diego de Urrea copied the first volume of a *Thesauro arauigo*[84] — possibly al-Jawharī's *Ṣiḥāḥ*. Later on, in the seventeenth century, eastern Christian scholars invited to Rome and Spain also brought copies of these works with them.[85]

Despite this evidence, there is comparatively less known about Arabic lexicography in sixteenth-century Spain than Arabic grammar. For this reason, when taken as a whole, Castillo's workbook is a remarkable testimony to the continuing tradition of classical Arabic lexicography in Spain during that century. Castillo had, of course, also studied grammar, and he was well aware how important his linguistic skill and accurate translation were for his personal and professional standing. Therefore, the materials he collected at this stage of his career were geared toward building new vocabularies and gaining experience in different registers.

Castillo's table of contents makes it quite clear that he envisioned his workbook being used as a lexicographical tool — whether for himself or future readers, fellow translators, or students, it is impossible to say. Although

---

80 On the study of Arabic in al-Andalus, including the teaching materials which circulated or were created, see Fierro, *Al-Andalus*, pp. 66–67.

81 Zanón Bayón, 'Los estudios de lengua árabe', p. 371.

82 The Biblioteca Nacional's copy of *Ṣiḥāh al-ʿarabiyya* was copied in 1505. Guillén Robles, *Catálogo de los manuscritos árabes existentes en la Biblioteca Nacional de Madrid*, doc. CXLIX. On Paéz de Castro and Hurtado de Mendoza's Arabic books, see Morata, 'Catálogo de los fondos árabes primitivos de El Escorial'.

83 Jones, 'Learning Arabic in Renaissance Europe (1505–1624)', p. 106.

84 Morata, 'Catálogo de los fondos árabes primitivos de El Escorial', p. 103.

85 García-Arenal and Rodríguez Mediano, *Un Oriente español*, p. 617.

the manuscript itself seems to have been composed at different times, and in a somewhat haphazard fashion, Castillo brought out its internal logic and organization as he collated it and provided a table of contents.[86] This table of contents included 'Lo Romançado por mi en la liga e trato de concordia con el xerife de parte suya con el rey D. phelipe nro senor' (That which was translated by me from Arabic into Spanish during the treaty negotiations with the Sharīf [Moroccan Sultan] on his behalf with our lord King Philip),[87] as well as an annotated list of 'Proverbs and sayings from al-Jawharī', and a long list of verses 'sacados de toda lectura moral arabiga con declaración de sus mas notables terminos para facilidad e comodo de la lección arabiga en las margenes' (taken from all of [my] readings of Arabic moral literature with a list in the margin of the most notable terms in order to facilitate the reading of the Arabic) which, from the text, seem to have been compiled over a lifetime of reading of what Castillo called 'karārīs min abwāb al-adab' (quires of the chapters of literary works).[88] Likewise, he included lists of legal and medical terms, annotated, and finally a copy of all of the Arabic inscriptions he had translated from the Alhambra and other notable Granadan monuments, again annotated. He also provided a long collection of Andalusi maxims, in 'lingue comunis', which he appears to have collected over many years.[89]

In addition to the materials that he listed in the index, Castillo found it useful to include a number of other lists and reference materials for different purposes in his workbook. These include copies of the letters concerning his professional tasks and appointment, especially regarding his salary.[90] He gave pride of place to a copy of the all-important 1582 *cédula* from Philip II naming him royal translator.[91] He also copied a number of other thematic vocabularies. These included Castillo's own word lists (discussed above), as well as more humanistic materials, like a copy of the sentences of Marcus Aurelius, and notes on a handful of Greek rhetorical terms.[92] Castillo kept several lists of significant vocabulary (*terminos obscuros y notables*), and copied lexical lists that were intended to help him in his new duties. Furthermore, he interpolated other thematic vocabularies, like the 'maneras de hablar cortesias a grandes principes' (the ways in which to speak courteously to great princes),

86 Madrid, BNE, MS 7453, fols 2^{r-v} of the initial unfoliated pages, read left to right.

87 Madrid, BNE, MS 7453, fols 1–71 in Castillo's foliation. This *cartulario* includes some of his correspondence with the royal secretary Gabriel de Zayas concerning the translation of the letters, and Castillo's account of the circumstances of translating each letter alongside Pedro de Deza, President of the Real Chancillería of Granada.

88 Madrid, BNE, MS 7453, fol. 208.

89 Madrid, BNE, MS 7453, fols 252^{b}–224^{b}, studied and edited by Corriente and Bouzineb, *Recopilación de refranes andalusíes de Alonso del Castillo*.

90 Madrid, BNE, MS 7453, fols 3^{v}–5^{v} in unfoliated pages beginning left and fols 15–17^{r} in Castillo's foliation on his report to Zayas and ensuing correspondence.

91 Madrid, BNE, MS 7453, fols 18^{v}–19^{r}.

92 On the early modern popularity of collected maxims, witticisms, and historical examples from the Latin humanist context, see Blair, *Too Much to Know*, pp. 298, 314.

including a chapter on the prayers (*min bāb al-duʿāʾ*) similar to those he was confronted by in Moroccan correspondence. He also had access to two of the most important works of Classical Arabic lexicography, drawing various 'examples' or maxims (*amthāl*) from the *Ṣiḥāḥ al-Jawharī* and copying four folios from al-Fīrūzābādī's *Qamūs*.[93] Indeed, all of these materials taken together give a very strong impression of being in line with the multi-generic lexical compendia that were supposed to teach certain elites (aspiring or otherwise) how to act in society throughout Renaissance Europe.

Thus, reading the manuscript workbook as a whole gives a sense of Castillo's professional learning curve and the materials that were available to him. Castillo's notebook was meant to serve him as a reference handbook for his professional work as a translator, and in particular for the high-stakes world of diplomatic service and international relations. Through this professional service, Castillo hoped to receive royal recognition (and a regular salary). It may have also served him as a reference for his role in composing the *plomos* forgeries — on fol. 108ʳ there is a passage which resembles the so-called 'Solomonic' writing of the Lead Books of Sacromonte — but, since those folios are undated, it is impossible to determine whether this was an exercise in copying similar calligraphy either before or after the books had been discovered. In the context of a handbook which seems highly geared to the needs of a professional chancellery translator, we may also suspect that these attempts were instead related to the Arabic cipher alphabets in which Castillo was also interested. Later (or earlier, if the workbook is read right to left) he provided a copy of 'Siete maneras que se hallan de alphabetos arabigos en manera de cifras' (Seven ways in which ciphered Arabic alphabets are found).[94] Of these alphabets, some characters resemble the ciphers described by al-Ḥajarī after his years of service in the Moroccan chancellery.[95] In addition to his catalogue of ciphers, his lexicon of princely courtesies, and his copious notes on vocabulary from the diplomatic correspondence, he included other materials in his workbook, such as the 'Nota de una carta arábiga bien notada' (Note about a well-formed Arabic letter).[96] This specimen, a general model rather than a copy of a specific letter, was one more example for his own future reference, in addition to those he had already translated in the context of Hispano-Moroccan diplomatic relations.

Each of the materials included by Castillo deserves further in-depth study. Nonetheless, collectively they reveal a great deal about the context in

93 The examples drawn from al-Jawharī are explanations of popular sayings, like 'كمستبضع تمر الى هجر' (the vendor of dates on the *hijra*, that is a deprecatory way of describing someone acting without sense, e.g., 'bringing coals to Newcastle'). These maxims were glossed in a similar manner to Castillo's marginal notes in his copies of the diplomatic correspondence.

94 Madrid, BNE, MS 7453, fols 220ᵛ–219ᵛ.

95 Mouline, *Le califat imaginaire d'Ahmad al-Mansur*, p. 226.

96 Madrid, BNE, MS 7453, fols 264ᵛ–63ᵛ.

which an Arabic expert conducted his professional life in sixteenth-century Spain. All of these materials served Castillo in his different obligations and ambitions to become the most senior Arabic expert in Spain. They also show that students of Arabic and professional Arabic experts in Spain were able to access a wide variety of materials upon which to generate the lexical expertise they needed for their very practical tasks. This included the most important examples of Classical Arabic lexicography, which would eventually become the foundation for the first printed works of European Arabic lexicography, including Erpenius, Golius, and Pococke.[97] Castillo was not a direct link in this particular chain of transmission, but the common reliance on these works of classical lexicography by the most famous representatives of Northern European orientalism and a *morisco* in the service of the Spanish state helps us to reframe our conventional narrative of Arabic learning in early modern Europe as being the preserve of missionaries and orientalists.

In seventeenth-century Spain, al-Jawharī's work — whether the same copy or a different version — was still available for consultation by a new generation of Arabists, like the Irish-born Jesuit Tomás de León.[98] Although there were fewer references to al-Fīrūzābādī's work (the three BNE versions are all eighteenth-century copies) until the arrival of Maronite scholars in the seventeenth century,[99] Castillo and his colleagues clearly had access to at least one small part of that widely disseminated work. As one final coincidence, or indication of a larger pattern yet to be discovered, Castillo's successor as Royal Translator, Diego de Urrea, would become involved in a long-term project of translating the *Qamūs* when he was working for Borromeo in Rome after 1616.[100]

Ennobling the Vernacular and Transmitting Courtly Registers in Early Modern Lexicography

The above inventory of Castillo's materials makes it clear, however, that his workbook was not only a manifestation of the continuity of use of Classical Arabic lexicography in *morisco* Spain, though its classical connections are highly significant. Rather, it was a composite handbook of materials, of a wide variety of genres and registers, from songs and proverbs in the vernacular 'linguae comunis' to various high registers of chancellery, poetic, and religious Arabic somewhere on the continuum between Middle Arabic and *fuṣḥā*. In short, although unrevised for a print publication, Castillo's manuscript portrays all the characteristics of a typical sixteenth-century European lexicographical

97 Haywood, *Arabic Lexicography*, pp. 122–23.

98 On this figure and his Vocabulista see Mediano, 'Fragmentos de Orientalismo español del s. xvii', pp. 272–73.

99 García-Arenal and Rodríguez Mediano, *Un Oriente español*, p. 617.

100 Floristán, 'Diego de Urrea (c. 1559-Octubre de 1616)', p. 249.

compendium of the type rolling off early presses for an increasingly demanding market of students, merchants, diplomats, and scholars.

The sixteenth century was the great age of multilingual lexicography in Europe, accompanying (and shaping) the so-called 'rise of the vernacular'. As Latin was 'dethroned' from its place as the European *lingua franca*, increasingly mobile Europeans had strong motives to acquire a working knowledge of other European vernaculars in order to conduct commercial, diplomatic or other activities as they travelled. Learning languages acquired a strong economic motive, and the growing print industry evolved into what would become a highly lucrative market, one already conditioned by a demand for manuscript language manuals in the fifteenth century.[101]

The language-learning manuals that dominated one portion of the print industry — notably in the publishing capital Antwerp, where there were incentives to learn the languages of the city's cosmopolitan inhabitants and networks — always included a targeted lexical component.[102] It was this turn to specialized vernacular lexicography, in addition to the new motives and materials, that distinguished the early modern language-learning tools in Europe from their medieval predecessors. The renaissance genres of *diálogos, colloquios*, and *nomenclaturas* were composite materials with a practical focus: theme-based vocabulary lists, practical dialogues or other sample texts (commercial letters, forms of address, prayers), and other key words a neophyte might need while interacting with speakers of other languages (chronological forms: days of the week, etc.). For comparison, in the workbook Castillo copied sample letters — both diplomatic and panegyric — names of the months of the Islamic calendar, prayers and forms of address for notables, and examples of Arabic cryptography.

The narrative of European vernacular lexicography is dominated by scholarly compilations on the one hand, and diplomatic and commercial motives on the other. When viewed from this perspective, Castillo's workbook — the composition of a Spaniard, after all — does not seem so foreign. Moreover, the formal characteristics of that type of genre are not dissimilar from the materials developed within that other language-learning narrative of the early modern period, the missionary and biblical-philological motives of orientalism. Thus, Castillo's workbook has a foot in both genres and both narratives, and is an example of why these distinct narratives — and the historiographies which depend upon them — should be in conversation.

Although he did not go so far as to provide sample dialogues, as commercial and humanist manuals did, Castillo's workbook shares a supremely practical character with the genre of vernacular lexicography. For example, in addition to the lists of legal, medical, cipher, and chronological terms, Castillo also

101 Hoock and Kaiser, 'Les manuels plurilingues à l'usage des marchands à l'époque moderne', pp. 71–72.

102 Acero Durántez, 'Lexicografía plurilingüe de los siglos XVI y XVII'.

preserved his translation of the safe conduct which had been issued to allow his friend Francisco Barredo, a *morisco* merchant, to trade in Morocco.[103] The language of al-Manṣūr's merchant passport is far more straightforward than his diplomatic letters, and did not require any special marginal annotations (the annotations in its margins concern the vocabulary found in the previous letter from al-Manṣūr to Philip). However, his interest in preserving an example of this kind of document is, in some ways, analogous to other kinds of formulaic documents provided to merchants as model specimens.

Castillo's linguistic expertise was simultaneously motivated by a deep personal interest in his various professional fields (actual and aspirational): medical, legal, religious, political (as an aspiring political councillor), diplomatic, and commercial. This mix of classical erudition and popular texts is also perhaps one of the most 'Spanish' features of the workbook since, as García-Arenal and Rodríguez Mediano have recently observed, 'to learn Arabic in sixteenth-century Spain was to attempt to bridge the gap between dialectal Arabic and written Arabic'.[104] The Andalusi poems and maxims (alongside their humanist analogue in the sentences copied from Marcus Aurelius) are a signal of the way Castillo mixed vernacular learning with classical erudition across traditions. Here it is worth emphasizing once again the continuous presence of Classical Arabic sources in this mix, especially lexicographical models like al-Jawharī and al-Fīrūzābādī, since these materials are more usually considered the foundation for later orientalism, and not a part of the *morisco* canon.

Castillo could have been exposed to the trends of vernacular lexicography and especially multilingual manuals in any number of sites, whether during his education at the University of Granada, in conversation with the courtly networks he aspired to join, which included humanists like Ambrosio de Morales; or in his various trips to the Escorial and his contact with scholars and officials there, including other multilingual specialists like Antonio Gracián, who eventually became Philip II's head translator.[105] He was certainly intimately involved in one particular branch of Spanish lexicography itself, the study of *Arabismos* in Spanish.[106] The didactic and referential nature of his workbook — while not identical to the formats developing in printed manuals — is similar enough to situate Castillo within a broader trend of lifelong language learners. Like the merchants who turned to language manuals to prepare for market exchanges, or the humanists who used

---

103 Madrid, BNE, MS 7453, fols 270^{r}–269^{r}.

104 García-Arenal and Rodríguez Mediano, 'Sacred History, Sacred Languages', p. 160.

105 Cabanelas, *El morisco granadino Alonso del Castillo*, pp. 124–30.

106 Castillo reviewed Francisco López Tamarid's *Compendio de algunos vocablos arauigos*, which was subsequently incorporated by Sebastian de Covarrubias into the *Tesoro de la lengua castellana* (1611). García-Arenal and Rodríguez Mediano, *The Orient in Spain*, pp. 126–27. On the role of other Granada Arabic translators in this phase of Spanish lexicography, see also Ruhstaller, 'Los inicios de la investigación sobre el arabismo léxico en español'.

linguistic references to complete editorial projects for the print market, he was certainly motivated to pursue new language skills to achieve his professional goals. In an early modern world where print, linguistic nationalism, and an ambivalence to the danger and promise of Arabic knowledge encouraged a diverse range of linguistic expertise in the servants of the Spanish state, Castillo's working habits and materials show that he was at the centre of more than one debate. Even when he was obliged to work from afar in Granada, and for years without the royal title he so ardently desired, Castillo lived far from on the margins.

Conclusion: Narratives from the Margins

Castillo's inclusion and mixing of various registers offers an alternative — though not incompatible — narrative about Arabic and Arabism to those of priestly, humanist, or crypto-Muslim Arabic study. In this alternative narrative, Arabic continued to be an important language of record and diplomacy for the Spanish state even at the end of the sixteenth century, and Arabic study bridged classical traditions and vernacular trends. Specialized Arabic expertise which relied on the classical tradition was available and cultivated by many agents of the monarchy, not only by scholars. Ultimately, the concomitant demand for Arabists and the repeated legal proscriptions of Arabic — which necessitated its own array of qualified experts to evaluate whether crime or heresy had taken place — generated a massive Arabic translation movement in Spain throughout the sixteenth century, which proved to be a balancing act between the interests of minority governance and Mediterranean empire. Nonetheless, in the midst of these many scales of ambivalence and competing agendas, professional Arabists continued to adapt the limits of their knowledge by seeking out a wide variety of materials.

Spanish Arabism was conditioned, like other early modern European orientalisms, by demands in different (although sometimes overlapping) cultural and political spheres, including biblical scholarship (and other humanist endeavours), medical (and other scientific) expertise, and commercial and diplomatic contact (including military conflict and conquest). Those who deployed this expertise were not only the priests, humanists, and orientalists who usually characterize the narrative of Spanish Arabism in the sixteenth century — but also *moriscos* like Castillo who spent a lifetime in the service of the state, embedded in administrative networks. Furthermore, the Spanish administration had an acute need for Arabic expertise in parallel to their diplomatic need for experts in French, Latin, Italian, and other vernaculars which would simultaneously inspire the founding of various state translation bureaus as the *Secretaria de interpretación de lenguas*.[107] These

107 Cáceres Würsig, 'Breve historia de la secretaría de interpretación de lenguas'.

language professionals also cultivated new knowledge through a combination of vernacular and classical materials, aspired to personal advancement and spent copious amounts of energy on professional rivalries and annoyances. Although Castillo, as a *morisco*, never achieved the social status of those royal secretaries, he would have been no stranger to their methods. Indeed, from the handwritten margins and between the lines of his notebook, a familiar story of curiosity and ambition emerges.

Bibliography

Manuscript Sources

Granada, Archivo Histórico Municipal de Granada, Libro de cabildo V (1556 y parte de 1557)

Madrid, Biblioteca Nacional de España, MS 7453

Primary Sources

Alonso del Castillo, 'Sumario e recopilación de todo lo romançado', in *Memorial histórico español: Colección de documentos, opúsculos y antigüedades que publica la Real Academia de La Historia*, Tomo III (Madrid: Real Academia de la Historia, 1852)

González, Bernardino, *Intérprete arábico: Epítome de la gramática arábiga*, ed. by Ramón Lourido Díaz (Madrid: Real Academia de la Historia, 2005)

Luis del Mármol Carvajal, *Historia del rebelion y castigo de los moriscos del reyno de Granada…* (Málaga: Por Iuan Rene, a costa del autor, 1600)

Miguel de Luna, *La verdadera hystoria del Rey Don Rodrigo: en la qual se trata la causa principal dela perdida de España y la cõquista que della hizo Miramamolin Almãçor Rey que fue del Africa, y de las Arabias* (Granada: Impr. por R. Rabut., 1592)

———, *Historia verdadera del Rey Don Rodrigo*, ed. by Luis Bernabé Pons (Granada: Editorial Universidad de Granada, 2001)

Secondary Studies

Acero Durántez, Isabel, 'Lexicografía plurilingüe de los siglos XVI y XVII', *Philologia Hispalensis*, 22 (2008), 7–36

Andrés, Gregorio, 'Los códices del Conde de Miranda en la Biblioteca Nacional', *Revista de Archivos, Bibliotecas y Museos*, 82.4 (1979), 611–27

Blair, Ann, *Too Much to Know: Managing Scholarly Information before the Modern Age* (New Haven: Yale University Press, 2010)

Cabanelas, Dario, 'El morisco granadino Alonso del Castillo, intérprete de Felipe II', *Miscelánea de estudios árabes y hebraicos. Sección Árabe-Islam*, 5 (1956), 19–42

———, 'El problema de Larache en tiempos de Felipe II', *Miscelánea de Estudios Árabes y Hebraicos. Sección Árabe-Islam*, 9 (1960), 19–53

———, *El morisco granadino Alonso del Castillo* (Granada: Patronato de la Alhambra, 1965; reprint 1991)

Cáceres Würsig, Ingrid, 'Breve historia de la secretaría de interpretación de lenguas', *Meta*, 49.3 (2004), 609–28

Castro, Francisco Vidal, 'Los diccionarios español-árabe: más de cinco siglos de lexicografía bilingüe', *Philologia Hispalensis*, 22 (2008), 319–45

Chorão Lavajo, Joaquim, 'Itinéraire Islamo-Arabisant de Clénard', in *Actas XVI Congreso de la Union Europea de Arabistas e Islamistas*, ed. by María Concepción Vázquez de Benito and Miguel Angel Manzano Rodríguez (Madrid: Agencia Espanola de Cooperacion Internacional, 1995), pp. 291–99

Corriente, Federico, 'La lexicografía árabe en España: Pasado, presente y futuro', in *Pensamiento y circulación de las ideas en el mediterráneo : El papel de la traducción*, ed. by Miguel Hernando de Larramendi Martínez (Cuenca: University of Castilla-La Mancha, 1997), pp. 133–49

Corriente, Federico, and H. Bouzineb, *Recopilación de refranes andalusíes de Alonso del Castillo* (Zaragoza: University of Zaragoza, 1994)

Cory, Stephen, *Reviving the Islamic Caliphate in Early Modern Morocco* (Farnham: Ashgate, 2013)

El Alaoui, Youssef, 'El jesuita Ignacio de las Casas y la defensa de la lengua árabe. Memorial al padre Cristobal de los Cobos, provincial de Castilla (1607)', *AREAS. Revista Internacional de Ciencias Sociales*, 30 (2011), 11–28

Espinar Moreno, Manuel, 'Escrituras árabes inéditas del siglo XV romanceadas por Alonso del Castillo', *Miscelánea de estudios árabes y hebraicos*, 46 (1997), 29–48

Feros, Antonio, 'Rhetorics of Expulsion', in *The Expulsion of the Moriscos from Spain: A Mediterranean Diaspora*, ed. by Mercedes García-Arenal Rodriquez and Gerard A. Wiegers (Leiden: Brill, 2014), pp. 60–101

Fierro, Maribel, *Al-Andalus: Saberes e intercambios culturales* (Barcelona: Icaria Editorial, 2001)

Floristán, José M., 'Diego de Urrea (c. 1559-Octubre de 1616), traductor de árabe, turco y persa en la corte de España: nuevas noticias biográficas', *Boletín de la Real Academia de la Historia*, 210.2 (2013), 227–73

Fournel-Guérin, Jacqueline, 'Le livre et la civilisation écrite dans la communauté morisque aragonaise (1540–1620)', *Mélanges de la Casa Velázquez*, 15 (1979), 241–59

García-Arenal, Mercedes, *Messianism and Puritanical Reform: Mahdīs of the Muslim West* (Leiden: Brill, 2006)

———, 'The Religious Identity of the Arabic Language and the Affair of the Lead Books of the Sacromonte of Granada', *Arabica*, 56.6 (2009), 495–528

García-Arenal, Mercedes, and Fernando Rodríguez Mediano, *Un Oriente español: los moriscos y el sacramonte en tiempos de contrarreforma*, Marcial Pons Historia (Madrid: Marcial Pons, 2010)

———, *The Orient in Spain: Converted Muslims, the Forged Lead Books of Granada, and the Rise of Orientalism*, trans. by Consuelo López-Morillas, Numen Book Series, CXLII (Leiden: Brill, 2013)

———, 'Sacred History, Sacred Languages: The Question of Arabic in Early Modern Spain', in *The Teaching and Learning of Arabic in Early Modern Europe*, ed. by Jan Loop, Alastair Hamilton, and Charles Burnett (Leiden: Brill, 2017), pp. 133–62

García-Arenal, Mercedes, Fernando Rodríguez Mediano, and Rachid El Hour, *Cartas marruecas: documentos de Marruecos en archivos españoles (siglos XVI–VXII)* (Madrid: Consejo Superior de Investigaciones Científicas, 2002)

Garrad, Keith, 'The Original Memorial of Don Francisco Nuñez Muley', *Atlante*, 2.4 (1954), 199–266

Gilbert, Claire, 'The Politics of Language in the Western Mediterranean c. 1492–c. 1669: Multilingual Institutions and the Status of Arabic in Early Modern Spain' (unpublished doctoral dissertation, UCLA, 2014), Available online at <https://escholarship.org/content/qt8w39b8gf/qt8w39b8gf.pdf> [accessed 03 March 2020]

———, 'The Circulation of Foreign News and the Construction of Imperial Ideals: The Spanish Translators of Aḥmad al-Manṣūr', *Memoria y Civilización*, 18 (2015), 37–70

———, *In Good Faith: Arabic Translation and Translators in Early Modern Spain* (Philadelphia: University of Pennsylvania Press, 2020)

Girard, Aurélien, 'Teaching and Learning Arabic in Early Modern Rome: Shaping a Missionary Language', in *The Teaching and Learning of Arabic in Early Modern Europe*, ed. by Jan Loop, Alastair Hamilton, and Charles Burnett (Leiden: Brill, 2017), pp. 189–212

Guillén Robles, Francisco, *Catálogo de los manuscritos árabes existentes en la Biblioteca Nacional de Madrid* (Madrid: Imprenta y Fundación de M. Tello, 1889)

Harvey, Leonard Patrick, *Muslims in Spain, 1500 to 1614* (Chicago: University of Chicago Press, 2005)

Haywood, John A., *Arabic Lexicography: Its History, and Its Place in the General History of Lexicography* (Leiden: Brill, 1960)

Hershenzon, Daniel, 'Traveling Libraries: The Arabic Manuscripts of Muley Zidan and the Escorial Library', *Journal of Early Modern History*, 18.6 (2014), 535–58

Hoock, Jochen, and Wolfgang Kaiser, 'Les manuels plurilingues à l'usage des marchands à l'époque moderne', in *Langues et langages du commerce en Méditéranée et en Europe à l'époque moderne*, ed. by Gilbert Buti, Michèle Janin-Thivos, and Olivier Raveux, (Aix-en-Provence: Presses Universitaires de Provence, 2013), pp. 71–82

Houston, Robert Allan, *Literacy in Early Modern Europe: Culture and Education, 1500–1880* (London: Longman, 2002)

Jones, John Robert, 'Learning Arabic in Renaissance Europe (1505–1624)' (unpublished doctoral thesis, University of London, 1988)

Koningsveld, Pieter Sjoerd van, 'Le parchemin et les livres de plombe de Grenade: Ecriture, langue, et origine d'une falsification', in *Nuevas aportaciones al conocimiento y estudio del Sacro Monte: IV Centenario de su fundación (1610–2010)*, ed. by María Julieta Vega García-Ferrer, María Luisa García Valverde, and Antonio López Carmona (Granada: Fundación Euroárabe, 2011), pp. 171–96

Koningsveld, Pieter Sjoerd van, and Gerard Wiegers, 'Apéndice documental', in *Nuevas aportaciones al conocimiento y estudio del Sacro Monte: IV Centenario de su fundación (1610–2010)*, ed. by María Julieta Vega García-Ferrer, María Luisa García Valverde, and Antonio López Carmona (Granada: Fundación Euroárabe, 2011), pp. 215–58.

Koningsveld, Pieter Sjoerd van, Gerard Wiegers, and Mónica Colominas Aparicio (trans.), *El Pergamino y los libros plúmbeos del Sacromonte. Edición crítica de los textos árabes y análisis de las ideas religiosas en ellos: Presentación de un proyecto de investigación holandés, Granada, 19 de marzo de 2019, 19.00–21.00. Con imágenes de los libros plúmbeos y del pergamino.* Rijswijk: Uitgeverij Avondrood. <https://dare.uva.nl/search?identifier = 946fc4f9-23db-4831-8391-0e0a8e5a2039> [accessed 16 June 2020]

Lentin, Jérôme, 'Middle Arabic', in *Encyclopedia of Arabic Language and Linguistics*, Managing Editors Online Edition: Lutz Edzard, Rudolf de Jong (Leiden: Brill, 2011), <http://dx.doi.org/10.1163/1570-6699_eall_EALL_COM_vol. 3_0213> [accessed 03 March 2020]

López-Baralt, Luce, and Reem Iverson, *'A zaga de tu huella'. La enseñanza de las lenguas semíticas en salamanca en tiempos de San Juan de La Cruz* (Madrid: Editorial Trotta, 2006)

López-Morillas, Consuelo, 'Language and Identity in Late Spanish Islam', *Hispanic Review*, 63.2 (1995), 193–210

Merino, Mercedes Abad, 'La traducción de cartas árabes en un pleito granadino del siglo XVI. El fenómeno del romanceado como acto judicial: Juan Rodríguez y Alonso del Castillo ante un mismo documento', *Al-Qanṭara*, 32.2 (2011), 481–518

Monroe, James T., *Islam and the Arabs in Spanish Scholarship (Sixteenth Century to the Present)* (Leiden: Brill, 1970)

Morata, Nemesio, 'Catálogo de los fondos árabes primitivos de El Escorial', *Al-Andalus*, 2.1 (1934), 87–182

Mouline, Nabil, *Le califat imaginaire d'Ahmad al-Mansur: Pouvoir et diplomatie au Maroc au XVIe siècle* (Paris: Presses Universitaires de France, 2009)

Muñoz, Nuria Martínez de Castilla, 'The Teaching and Learning of Arabic in Salamanca in the Early Modern Period', in *The Teaching and Learning of Arabic in Early Modern Europe*, ed. by Jan Loop, Alastair Hamilton, and Charles Burnett, The History of Oriental Studies, III (Leiden: Brill, 2017), pp. 163–88

Narváez-Córdova, María Teresa, 'Writing without Borders: Textual Hybridity in the Works of the Mancebo de Arévalo', *Medieval Encounters*, 12.3 (2006), 487–97

Pallarés, Eugenio Ciscar, '"Algarabía" y "algemía": Precisiones sobre la lengua de los moriscos en el Reino de Valencia', *Al-Qantara*, 15.1 (1994), 131–62

Pedraza, Amalia García, *Actitudes ante la muerte en la Granada del siglo XVI : Los moriscos que quisieron salvarse* (Granada: Editorial Universidad de Granada, 2002)

Rodríguez Mediano, Fernando, 'Fragmentos de Orientalismo español del s. xvii', *Hispania*, 66.222 (2006), 243–76

Ruhstaller, Stefan, 'Los inicios de la investigación sobre el arabismo léxico en español', *Bulletin hispanique. Université Michel de Montaigne Bordeaux*, 115.1 (2013), 253–70

Starczewska, Katarzyna Krystyna, and Mercedes García-Arenal, '"The Law of Abraham the Catholic": Juan Gabriel as Qur'ān Translator for Martín de Figuerola and Egidio de Viterbo', *Al-Qantara*, 35.2 (2014), 409–59

Szpiech, Ryan, *Conversion and Narrative: Reading and Religious Authority in Medieval Polemic* (Philadelphia: University of Pennsylvania Press, 2013)

Vincent, Bernard, 'Reflexión documentada sobre el uso del árabe y de las lenguas románicas en la España de los moriscos (ss. XVI–XVII)', *Sharq Al-Andalus*, 10–11 (1993), 732–48

Wiegers, Gerard, *Islamic Literature in Spanish and Aljamiado: Yça of Segovia (Fl. 1450), His Antecedents and Successors* (Leiden: Brill, 1994)

———, 'El contenido de los textos árabes de los plomos: el *Libro de los misterios enormes* (*Kitāb al-Asrār al-'Azīma*) como polémica islámica anticristiana y antijudía', in *Nuevas aportaciones al conocimiento y estudio del Sacro Monte: IV Centenario de su fundación (1610–2010)*, ed. by María Julieta Vega García-Ferrer, María Luisa García Valverde, and Antonio López Carmona (Granada: Fundación Euroárabe, 2011), pp. 197–213

Zanón Bayón, Jesús, 'Los estudios de lengua árabe entre los moriscos aragoneses a través de los manuscritos de la Junta', *Sharq Al-Andalus*, 12 (1995), 363–74

PIETRO DANIEL OMODEO

Johannes Regiomontanus and Erasmus Reinhold

Shifting Perspectives on the History of Astronomy

In 1549 the Wittenberg professor of mathematics Erasmus Reinhold published a new edition of Johannes Regiomontanus's *Oratio… quam habuit ipse Patavii in praelectione Alfarghani* (Oration… he held in Padua in his Lectures on Alfarghani). It was a eulogy of mathematics that Regiomontanus had delivered in 1464 at the University of Padua. The text was already available in print; it had appeared under the title *Oratio introductoria in omnes scientias mathematicas* (Introductory oration to all mathematical sciences) in Nuremberg in 1537, under the auspices of the learned Lutheran humanist, Philipp Melanchthon. There are clear ties between Reinhold and Melanchthon, as well as between the two editions: Melanchthon had supported Reinhold's appointment as a professor of mathematics at the University of Wittenberg in 1536 and had inspired a humanistic approach to mathematics at that university. The 1549 reprint of Regiomontanus's oration was preceded by an oration that Reinhold had held on the official conferment of the Master of Arts degree to eight students at Wittenberg. The oration celebrated Regiomontanus as a model scholar, as the title openly indicates:

> Oratio de Johanne Regiomontano mathematico, in Renunciatione gradus Magisterii Philosophici, recitata ab Erasmo Rheinholt Salveldensi Mathematum professore
>
> (Oration on Johannes Regiomontanus, mathematician, pronounced by the professor of mathematics Erasmus Reinhold of Saalfeld on the occasion of the conferment of the title of Master of Arts) (Wittenberg, 1549)

Pietro D. Omodeo • (pietrodaniel.omodeo@unive.it) is a professor of historical epistemology at Ca' Foscari University of Venice and a permanent guest of the Max Planck Institute for the History of Science, Berlin. His research and teaching focuses on early modern science and political epistemology. Among his recent works, he published *Political Epistemology: The Problem of Ideology in Science Studies* (Dordrecht: Springer, 2019). He is the Principal Investigator of the ERC project 'EarlyModernCosmology' (GA 725883). The essay in this volume is a contribution to this project.

Premodern Translation: Comparative Approaches to Cross-Cultural Transformations, ed. by Sonja Brentjes and Alexander Fidora, Contact and Transmission, 2 (Turnhout: Brepols, 2021), pp. 165–186

10.1484/M.CAT-EB.5.122137

This text was later included in Melanchthon's *Selectissimae orationes* (Strictly selected orations) (1551). This publication in a collection of works by Melanchthon makes us wonder whether Reinhold or Melanchthon actually composed the oration. In any case, it can be seen as the product of collaboration between the leading Wittenberg intellectual, Melanchthon, and the distinguished professor of mathematics, Reinhold.[1]

The orations by Regiomontanus and Reinhold offer interrelated but different perspectives on the history of mathematical astronomy. Regiomontanus was one of the most reputed western European mathematicians and astronomers of the generation before Copernicus, profoundly influenced by the culture of the Italian Renaissance.[2] The second scholar, Reinhold, has been extolled with good reason as 'the leading astronomer of the sixteenth century' with regard to computational astronomy.[3] His astronomical tables, the *Prutenicae tabulae* (Prussian Tables) (Tübingen, 1551), were well-received and widely disseminated among specialists as a replacement for the pre-existing tables, the *Alfonsine*, up to the publication of Kepler's *Rudolphine Tables* in the seventeenth century. Reinhold computed his tables using Nicolaus Copernicus's parameters and used *De revolutionibus orbium coelestium* (On the Revolutions of the Celestial Orbs) (Nuremberg, 1543) as a basis for his teaching of planetary modelling, which paved the way for Tycho Brahe's geo-heliocentric hypotheses.[4] His work was embedded within the Wittenberg academic environment during the period of the cultural consolidation of the Lutheran Reform.

In this essay, I discuss the journey from Regiomontanus's historical perspective to Reinhold's and point out the continuities and discontinuities between these two protagonists of early modern mathematical astronomy. My inquiry aims to cast light on the transformations of the astronomical culture in the century or so from the scientific activity of the former to the latter, moving from Renaissance humanism to the scientific culture of the Reformation. Furthermore, my discussion permits us to observe a shift in narratives about transfer and translation that underlies how these authors dealt with the history of their discipline.

---

1 On the close collaboration between Melanchthon and his pupils, see Scheible, *Melanchthon in seinen Schülern*, and especially Thüringer, 'Paul Eber (1511–1569)'. For the sake of brevity, in this essay I will refer to it simply as Reinhold's oration, although its possible alternative or double attributions should be kept in mind.

2 Regiomontanus's inscription in the culture of Italian classicist humanism was stressed, even exaggerated, by Rose, *The Italian Renaissance of Mathematics*, chapter 4.

3 Gingerich, 'Reinhold, Erasmus', p. 365.

4 Reinhold's own commentary of Copernicus, *Commentarius in opus Revolutionum Copernici*, is preserved as a manuscript in Berlin at the Staatsbibliothek, collocation MS lat. fol. 391. For a transcription, see Copernicus, *Gesamtausgabe (Receptio Copernicana)*, pp. 189–358.

Padua, 1464: A Humanistic Perspective on the History of Astronomy

Many historians of science have discussed Regiomontanus's 1464 *Oratio introductoria in omnes scientias mathematicas* at length. Robert Goulding defined it as 'the first modern history of mathematics', since in it, 'the names of historical figures far outnumber the legendary progenitors of the arts'. In it, Regiomontanus stressed the vitality of the discipline with reference to modern authors.[5] Paul L. Rose saw Regiomontanus's oration as an example of what he deemed the 'Italian Renaissance of mathematics'. It is indeed a typical product of the fifteenth-century humanistic culture. However, in contrast to earlier interpretations that pit Renaissance literacy against Scholastic education in a rather schematic manner, recent assessments caution against one-sided interpretations that regard the work as an anti-Scholastic treatment of mathematics that opposed curricular university teaching. Rather, the two poles of learned Renaissance science, that is, the philological-classicist and the Scholastic, are merged in Regiomontanus's oration.[6] Moreover, this text reveals the conceptions of culturally hegemonic elites, given Regiomontanus's participation in the most influential humanistic milieus of Italy and his academic pedigree as a university-trained scholar. As the historian of mathematics Moritz Cantor stated, 'the importance [of the *Oratio introductoria*] especially rests on its casting a bright light on Regiomontanus's mathematical knowledge and on widespread historical opinions of his time'.[7] Several reliable studies placing Regiomontanus's oration in the context of his life and work already exist; the most significant among them by Ernst Zinner and Michela Malpangotto. Therefore, I limit my analysis to two aspects that I deem to be particularly relevant to an assessment of his historical outlook:[8] first, the rhetorical codification of his discourse and second, his understanding of culture, history, and transfer in light of Rose's suggestion that 'translation is [...] one of the pillars of Regiomontanus's concept of a mathematical renaissance'.[9]

---

5 Goulding, *Defending Hypatia*, pp. 8–10.

6 Cf. Byrne, 'A Humanist History of Mathematics?'.

7 Cantor, *Vorlesungen über Geschichte der Mathematik*, p. 238: 'ihre Wichtigkeit liegt insbesondere darin, dass sie auf das mathematische Wissen Regiomontanus und die damals verbreiteten geschichtlichen Meinungen ein helles Licht wirft'.

8 Zinner's original German monograph (1968) is available in English. For the oration, see Zinner, *Regiomontanus*, pp. 69–74. See also Malpangotto, *Regiomontano e il rinnovamento del sapere matematico e astronomico nel Quattrocento*, pp. 42–43, 75–94. Malpangotto also offers an annotated transcription, see Malpangotto, *Regiomontano e il rinnovamento del sapere matematico e astronomico nel Quattrocento*, pp. 133–46.

9 Rose, *The Italian Renaissance of Mathematics*, p. 96.

The Genre of Regiomontanus's *Oratio*

Regiomontanus's oration belongs to the rhetorical genre of eulogies of the arts. As has been recently pointed out in comparative analyses of Renaissance *laudationes* of astronomy and mathematics, these types of speeches were often given during official public occasions in the presence of university scholars, patrons, or academicians. They also functioned as a way of introducing scientific works. Sources as diverse as Girolamo Fracastoro's introduction to his *Homocentrica* (Venice, 1538), Reinhold's dedicatory letter in his edition of Ptolemy's *Almagest*, Book I (Wittenberg, 1549), Clavius's preface to his commentary of Sacrobosco's *De sphaera* and the public speech on mathematics delivered by the Scottish mathematician Duncan Liddel at Helmstedt in 1591, all exhibit the same structural patterns and resort to established 'rhetorical commonplaces'.[10] Such topoi include the nobility of astronomy as descending from the divinity of its subject, the heavens, and from the certainty of its geometrical demonstrations. Further instances of stock narratives are the Egyptian origin of geometry that derived from the need for accurate land measurement mechanisms after the Nile inundations (a story ultimately derived from Herodotus via Isidore of Seville) or references to Archimedes' inventions to illustrate the usefulness of mathematics. As far as the structure of eulogies of arts is concerned, they generally followed a common trajectory: they opened with a preliminary definition of the discipline — often including a clarification of the meaning of its name and its location in the system of sciences — followed by a historical overview stressing the antiquity of the discipline concerned, the celebration of its usefulness (*utilitas*) and finally, its intrinsic nobility (*dignitas*).

Regiomontanus's oration follows this standard construction. He begins with a *captatio benevolentiae*, addressing his audience of illustrious professors and students at the University of Padua. He then provides an etymological explanation of the name of *mathematica*. 'We' the Latins, Regiomontanus argues, derived this word from the Greek verb μανθάνω, which means *disco* (to learn). Hence, μαθηματική corresponds to the Latin *disciplina* (discipline). Regiomontanus proceeds to address the most general and abstract arts of the quadrivium —geometry and arithmetic, dealing with continuous and discrete quantities, respectively. Regiomontanus identifies their origins and lists the most important works on the subject from antiquity and the Middle Ages, accurately indicating which ones have been translated from Greek and Arabic into 'our language', that is, Latin. After 'the primary disciplines,

10 I expand on these topics in Omodeo, 'Utilitas Astronomiae in the Renaissance', dealing specifically with Liddel's oration in Omodeo, 'Liddel's Oratio de Praestantia Mathematicarum'. As for commonplaces, recurring themes can be referred to as 'rhetorical commonplaces', insofar as they are derived from stock ideas and narratives employed by Renaissance authors in their speeches and texts. See Hotson, *Commonplace Learning*.

which contemplate quantity in an absolute manner' (*disciplinae primariae, quae videlicet absolutam contemplantur quantitatem*), Regiomontanus tackles the 'secondary' disciplines. As he writes, he 'descends' (*descendo*) from the loftier disciplines to the subaltern, those dealing with 'concrete' quantities (*quantitates contracto quodam modo*). They include astronomy, defined as the science of movable magnitudes, music, on acoustic numbers (*numeri sonori*) and perspective, on radial lines. Here, he also mentions additional *scientiae mixtae*: the science of weights (*scientia de ponderibus*), aqueducts, and the treatment of speed by means of proportions (*de proportione velocitatum in motibus*). Regiomontanus additionally expands on astronomy. He regards it as the most excellent of all mathematical disciplines, not only in comparison to the subaltern ones but even in contrast to geometry and arithmetic, and also considers it the most ancient. In fact, as it related to the world in its entirety, he argues that astronomy is as ancient as Creation (*mundo concreata*). Regiomontanus offers an overview of the main classical and medieval works on astronomy and to a lesser extent, on music and optics (or perspective). This is followed by a section on the *utilitas* of mathematics and a disquisition on their *dignitas*. Dignity, according to Regiomontanus, is indicated through the strength and certainty of mathematical demonstrations and truths superior to those of philosophers. Furthermore, mathematics has propaedeutic import, as it is necessary for furthering philosophical studies (and particularly, the understanding of many Aristotelian works and passages), as well as for advancing studies in the higher faculties of medicine, jurisprudence, and theology. Concerning the theological and pious relevance of mathematical disciplines, Regiomontanus stresses that astrology specifically deals with God's Providence. This is a theme that would become central to the astronomical culture of Melanchthonian universities.[11]

Language and Knowledge Acquisition

Language plays a central role in Regiomontanus's conceptual map of culture and in his understanding of scientific transfer by means of translation. In the oration, one of the first issues he addresses is the renewal of the humanities as the basis for the revitalization of mathematical sciences. He also stresses the limitations of language for expressing mathematical truths. In fact, the excellence of mathematics seems to escape the communicative potentiality of *any* language, no matter how outstanding it is:

> Sed per Deum immortalem quis unquam satis accurate tam speciosas Nimphas exornare poterit? Cum Platonis in dicendo suavitas, tum *Ciceronis*

11 Sachiko Kusukawa has pointed out Melanchthon's providential interpretation of astrology Kusukawa, *The Transformation of Natural Philosophy*; cf. Brosseder, *Im Bann der Sterne*.

nostri lingua disertissima non adsint, qui etsi reviviscent haud quaquam pro dignitate idipsum assequerentur.

> (But for God Immortal! Who has ever been capable of adorning such splendid Nymphs [that is, the mathematical arts]? Not even Plato's suavity of expression or *our* Cicero's superior linguistic eloquence would be of help. Even if they came back to life, they would by no means have more chance of attaining this goal.)[12]

Cicero is the first author Regiomontanus addresses as 'noster' (ours, i.e., one of the authors who wrote in our language) followed by Boethius (*Boecius noster*) and Witelo (*Vitellio autem noster*). They are also indicated as *nostri*, a connotation evidently used to refer to scholars who composed their works in Latin. Hence, language defines Regiomontanus's cultural community. He calls his cultural community *Latinitas* (Latinity), a universal essence parallel to other language and cultural communities, in particular Greek and Arabic. Languages ideally inform and circumscribe nations (*nationes*), which in turn are subdivided into smaller-scale entities. The community of *Latinitas* subsumes many nations, such as those of the *Germani*, *Itali*, *Galli*, and *Britones*.

In this conceptual framework, intellectual history is the entangled unwinding of science along a plurality of cultural developments and across linguistic communities — the Latins, the Greeks, the Arabs, but also the Persians (*Persae*) and the Indians (*Indi*) — whose exchanges occur through appropriation, elaboration, and new syntheses. Euclid, for one, was able to produce a synthesis of all geometrical *conclusiones* (theses) produced by the generations preceding him since the Egyptians had created the discipline. Similarly, Ptolemy perfected astronomy, whose origins, as already mentioned, are traced back to the world's creation. Plato of Tivoli expressed a similar understanding in his thirteenth-century translation of Albategnius's (al-Battani's) *De motu stellarum* (The Motion of the Stars). Regiomontanus extolled the superiority of Albategnius, the Arabic follower of Ptolemy, over his Hellenistic master for the brevity and clarity of his synthesis, which made his work more suitable for translation than the *Almagest* itself:

> Cumque eius [Ptolemaei] imitatorem perfectum inter Arabes et Albategnium depraehenderem, quique Ptolemaei prolixitatem compendiosae coartat eiusque errores emendans, quae quidem rarissimi sunt, non ipsi sed Abrachis [Hipparchis] radici imputat [...] Hunc inquam Albategnium meo labore Deo propiciante trasferendum, et latinis auribus offerendum censui.
>
> (I found that Albategnius is a perfect imitator [of Ptolemy] among the Arabs. He synthesizes Ptolemy's prolixity while correcting his errors, which are actually very few and are attributed not to himself but

12 Johannes Regiomontanus, *Oratio introductoria in omnes scientias mathematicas*, quoted from *Oratio Iohannis de Monteregio quam habuit ipse Patavii in praelectione Alfragani* in Melanchthon, *Selectissimae orationes*, fol. 173^{v} (italics mine); my translation.

> rather to Hipparchus. [...] Hence, I decided to translate [*transferre*] the aforementioned Albategnius with my work and the help of God, and to offer him to Latin readers [literally, 'ears'].)[13]

Arguably, Peuerbach and Regiomontanus's most celebrated work, *Epytoma in Almagestum Ptolemaei* (Epitome of Ptolemy's *Almagest*) (1496), had the same aim as Albategnius's work, that is, to offer an improved synthesis of the *Almagest*.

According to Regiomontanus, translation is crucial for scientific appropriation and refinement, especially for those — the Latins — who accessed knowledge in delay. Plato of Tivoli, whose work Regiomontanus knew and appreciated, stressed the scientific inferiority of the Latins relative to other linguistic communities:

> In bellorum quidem foelicitate in imperii dilatatione Roma, non solum Aegyptum et Graeciam, verum omnis quotquot sunt nationes antecessit. In artium vero gymasiis, in disciplinarum speculationibus, licet [...] non tantum Aegypto vel Graecia, sed et Arabia longe inferior extitit.
>
> (Rome certainly surpassed not only Egypt and Greece but also all other existing nations as far as warfare and the capacity to extend its dominion are concerned. However, in the exercise of the arts and in scientific speculations it was by far inferior not only compared to Egypt and Greece but also to Arabia.)[14]

In the mid-fifteenth century, Regiomontanus's evaluation of the state of the art of the sciences is less desperate. He sees the process of translation and acquisition of works from other cultures as more advanced. He lists the most important acquisitions, as well as the classics that were not yet available to the Latinity.[15] Nonetheless, the idea of a scientific delay of the Latin world is as present in Regiomontanus's reflection as it is in his predecessor's assessment, that is, in Plato of Tivoli. Astronomy is a case in point:

> Ad Latinos postremo venitur [astronomia], qui munera huius disciplinae celeberrima a praefatis nationibus suscepere.
>
> (Eventually, [astronomy] came to the Latins, who received the most celebrated accomplishments of the aforementioned nations [Greeks, Arabs, Indians and Persians].)[16]

---

13 *Praefatio Platonis Tiburtini in Albategnium*, in Alfraganus, *Rudimenta astronomica Alfragani item Albategnius*; my translation.

14 *Praefatio Platonis Tiburtini in Albategnium*, in Alfraganus, *Rudimenta astronomica Alfragani item Albategnius*; my translation.

15 In the *Oratio introductoria*, quoted in Philipp Melanchthon, *Selectissimae orationes*, fol. 179^{v}, Regiomontanus notes, for instance, that 'Arabes praeterea quantum in hoc genere artium [astronomia] valuerunt, testes ostendunt Albategnius, quem Latinum fecit Plato quidam Tiburtinus; itam Geber Hispalensis Gerardo quodam Cremonensi traductus, quem Albertus Magnus in Speculo astronomiae correctorem Ptolemaei vocare non formidat [...]'.

16 Johannes Regiomontanus, quoted in Philipp Melanchthon, *Selectissimae orationes*, fol. 180^{r}; my translation.

According to Regiomontanus, this field of study was newly glorified in the Latin West by scholars such as Albertus Magnus (1206–1280), Pietro d'Abano (1257–*c.* 1315), Giovanni Dondi dall'Orologio (*c.* 1330–1388) and finally, by German scholars, beginning with Regiomontanus's master, Georg Peuerbach (1423–1461).[17]

Regiomontanus's Historical Perspective

In Regiomontanus's eyes, the origin of the sciences is practical. As he remarked in his preface to Euclid's *Elements*, 'geometry, like all other disciplines, is established first in practice [*in usu*] then in theory [*in arte*]'.[18] Regiomontanus traces the origins of geometry back to a political decision made by an Egyptian ruler (*princeps*), who devised a fair means of measurement capable of dispelling controversies over land possession and boundaries after recurring inundations of the Nile.[19] Regiomontanus elaborates on the practical dimension of the mathematical arts in the section of his oration dedicated to the theme of the *utilitas*. He first mentions the mechanical arts. In his *Preface* to the *Elements*, he applies Hugh of Saint Victor's (*c.* 1096–1141) classifications. These are *lanificium* (wool weaving), *armarua* (construction of weapons), *agricultura* (agriculture), *navigatio* (navigation), *venatio* (hunting), *medicina* (medicine), and *theatria* (theatrical arts). Computation is fundamental for commerce and it is subsequently advanced and perfected by merchants (*negociatores*) through its constant use. Geometry is useful for war, for ballistic as well as for civil applications, such as the construction of sundials and musical instruments. Civil endeavours and activities linked to technology, transportation, and architecture require mathematics as well. Incorrect computations can have ruinous consequences on construction and engineering projects. Regiomontanus uses the example of a mathematical advisor who vetoed a project to connect the Red Sea to the Mediterranean.[20]

The Greeks play a prominent role in Regiomontanus's history of mathematics. Authors from classical and Hellenistic Antiquity are mentioned with

---

17 Johannes Regiomontanus in Philipp Melanchthon, *Selectissimae orationes*, fol. 181^{r}: 'Quocirca post plurimos nationis nostrae insignissimos Astrorum contemplatores, te tandem Georgi de Peuerbach Germanorum perenne decus appellare iubeor'.

18 Johannes Regiomontanus, *In Elementa Euclidis Praefatio*, in Alfraganus, *Rudimenta astronomica Alfragani item Albategnius*: 'Geometria, sicut et reliquae discipline, prius praecessit in usu quam in arte'; my translation.

19 Johannes Regiomontanus, *Oratio introductoria*, in Philipp Melanchthon, *Selectissimae orationes*, fols 175^{r-v}.

20 Johannes Regiomontanus, in Philipp Melanchthon, *Selectissimae orationes*, fol. 184^{r}: 'Sinum Arabicum, vocant mare rubrum, Rex quidam mari nostro mediterraneo continuare tentasset, nisi geometra monitore totam Aegyptum aquarum illuvie submersum iri didicisset'; my translation.

admiration and their availability in Latin translations is correspondingly indicated: Euclid was translated (the verb used here is *verto*) by Boethius and his works were commented on during the Middle Ages by many; Apollonius's eight books on conic sections have not yet appeared in Latin (*nondum vidit latinitas*); Archimedes' works have been only partially translated into Latin (*Cremonensis Latinum ex Graeco reddidit*); Diophantus's work is not available in Latin as of yet (*nemo usque hac ex graecis latinos fecit*); and so on and so forth. Modern Greeks are only alluded to. After mentioning the lack of any translation of Theon's commentary of the *Almagest*, Regiomontanus tells his audience: 'I will pass in silence many Greek professors of this art, to avoid boring you'.[21] In general, the acquisition of Greek sources, their transmission and translation into Latin are indicated as an ongoing task. Regiomontanus saw his activity in Rome following that of Cardinal Bessarion: as a contribution to a larger project of transfer and appropriation of Greek sources and scholarship.[22]

The Arabs have a prominent place in Regiomontanus's history as well. Their proficiency in arithmetic is exemplified through the use of Arabic expressions such as 'algebra' ([*disciplina*] *quam hodie vocant Algebram Arabico nomine*). Albategnius, Geber, and Alfarghani [*Alfraganus*] are clear examples of excellent astronomical scholars, along with Alhazen, who was distinguished in the field of optics. Avicenna and Averroes are also mentioned in the part of the oration that deals with the relevance of mathematics for philosophy and medicine. Averroes is simply indicated as *commentator* (the commentator of Aristotle). Regiomontanus particularly praises his attempts to develop astronomy using concentric spheres according to the requirements of Aristotle's *Metaphysics* XII.[23]

---

21 Regiomontanus in Philipp Melanchthon, *Selectissimae orationes*, fol. 179v: 'Multos denique huius artis Graecos professores silentio praetereundos censeo, ne fastidio afficiamini'; my translation.

22 Regiomontanus in Philipp Melanchthon, *Selectissimae orationes*, fol. 182r: 'Duce itaque patrono communi in Romam profectus, more meo literis exerceor, ubi scripta plurima Graecorum clarissimorum ad literas suas discendas me invitant, quo Latinitas in studiis praesertim Mathematicis locupletior redderetur'.

23 Regiomontanus in Philipp Melanchthon, *Selectissimae orationes*, fol. 186v: 'Quantam gloriam commentator adeptum se putavit, si Astronomiam concentricam tradidisset, testimonio suo docemur, qui totam ferme aetatem in ea re consumens desperasse tandem se confitetur'. Regiomontanus's adherence to an Aristotelian cosmology of spheres has been pointed out by Shank, 'Regiomontanus on Ptolemy, Physical Orbs, and Astronomical Fictionalism'. The relevance of Arabic science for the advancement of Latin science is further revealed by Regiomontanus's indication of the Arabic mediation of Euclid's work. In his preface to Euclid's *Elements*, he remarks: 'Titulus est [...] Primus liber Euclidis philosophi de arte Geometria incipit. Vel, Incipit ars Geometriae, continens 364. Propositiones, ab Euclide in Arabico compositae, et ab Adelhardo Goth. in Latinum assumpta'. (Johannes Regiomontanus, *In Elementa Euclidis Praefatio*).

Appropriating Classical Sources

In the Padua oration, Regiomontanus offered a comprehensive overview of the history of mathematical sciences and the state of the art.[24] The last part entailed a lament about the neglect of the mathematical disciplines in university curricula. At universities, as he remarked, any amateur could claim to be an astronomer based only on a superficial knowledge of Sacrobosco (*prima rudimenta Iohannis de Sacro Busto*) and the most elementary part of Alkindi's work. By teaching classes based on Alfarghani, Regiomontanus intended to contribute to the restoration (*reparare*) of astronomy, a discipline that he deemed to be extremely useful for philosophy, medicine, jurisprudence, and theology. 'What now remains to be done — these are the concluding words of the oration — is to welcome all together Alfarghani, the illustrious historian, at the entrance of [our] home'.[25]

This plan for a scientific restoration most clearly emerges in Regiomontanus's dedication to Bessarion, of the *Epytoma*, wherein he emphasized the virtue of the ancients and the decadence of the present times.

> Admiranti mihi saepenumero, vel potius graviter et inique ferenti, tam raros esse aetate nostra optimarum disciplinarum non modo praeceptores, verum etiam studiosos, satis compertum videtur depravata potius hominum natura id fieri, ut ad vitia proclives virtutem ac bonas artes pro nihilo habeant, quam ut rerum ipsarum difficultas eos abterreat.
>
> (I have often wondered, or, rather, I have been forced to bitterly acknowledge, that, in our age, there are very few teachers of the most illustrious disciplines — and even very few learners! It seems to me quite evident that the inclination to vices alongside the contempt of virtue and the good arts depends on human nature rather than on the fear of the matter's difficulty.)[26]

24 The beginning of Regiomontanus's oration here follows *Oratio introductoria*, in Melanchthon, *Selectissimae orationes*, fol. 174^{r}: 'Memorare possem in primis originem nostrarum artium, et apud quas gentes primum coli coeperint, quo pacto ex linguis peregrinis variis ad Latinos tandem pervenerint, qui in hisce disciplinis apud maiores nostros claruerunt, et quibus nostra tempestate mortalibus palma tribuitur'.

25 Johannes Regiomontanus, in Philipp Melanchthon, *Selectissimae orationes*, fol. 190^{r}: 'Nunc reliquum est Alfraganum insignem Astronomiae historicum ad limina domus uno verbo salutemus'. At another point, Alfarghani is mentioned as the author of the *historia* of astronomy; cf. Johannes Regiomontanus, in Philipp Melanchthon, *Selectissimae orationes*, fol. 174^{r}: '[…] primordia Alfragani, qui nudam totius Astronomiae tradit historiam'. Here, history simply refers to a presentation of facts, of the so-called *quia*, and not of the profound causes of natural phenomena, the so-called *propter quid*. Throughout the sixteenth and seventeenth centuries, *historia* would be used to refer to literature of observational reports, for instance on comets or Galileo's *Istoria e dimostrazioni intorno alle macchie solari* (1613), dealing with telescopic wonders. For a recent discussion, cf. Jardine, 'Rothmann's Dialexis on the Comet of 1585'.

26 Johannes Regiomontanus, *Epytoma in Almagestum Ptolomei*, fol. a2^{r}; my translation.

Here the combination of the terms *virtus* and *bonas artes* is a hendiadys. The value of the *bonae literae* — of literature and science that opposed the philosophical confusion that reigned in the university disputes — is linked to the myth of a rebirth of pristine splendour. Inspired by the humanistic endeavours to recover the spirit of the classics, Regiomontanus looks back to those he designates as the *maiores nostri*. According to him, the time of *our* ancestors was not yet corrupted by ambition and other immoderate desires (*cupiditates*). In the golden age of classicism 'only virtue was praised and everybody was satisfied with his own' (*sola virtus in precio erat: sua cuique satis placebat*). In contrast, the present times are dominated by the desire for wealth, which discredits the 'good arts': 'nothing is appreciated here, apart from gold, while the arts are run down' (*hinc nihil preter aurum suave creditum est; disciplinae probro habite*). In the Latin world, the wisdom of the ancients was lost due to laziness and vice:

> Haec causa igitur est cur pauci aetate nostra docti sint, cur pauci studiosi, cur iaceant studia bonarum artium et quasi sepultae emergere ac suscitari non possint
>
> > (Therefore, learned people and learners are few in our time and the study of the good arts is neglected so that they can hardly emerge or be resuscitated from their graves.)[27]

After his lament on the present state of affairs, Regiomontanus praises astronomy, which lifts both the eyes and the spirit to the heavens, realizing the natural aspiration of human beings to reach that which is lofty and divine. Regiomontanus clearly draws upon and continues a classical theme. Following Plato's *Timaeus* (47 BC) and Cicero's *De natura deorum* (II 56 140) and anticipating Pico's *De hominis dignitate oratio* (Oration on the Dignity of Man, composed in 1486), Regiomontanus asserts that 'God the Astrologer' (*divinum Astrologiae numen*) created human beings to have an upright position so that they could contemplate the heavens.[28]

---

27 Johannes Regiomontanus, *Epytoma in Almagestum Ptolomei*, fol. a2r; my translation.

28 Johannes Regiomontanus, *Oratio introductoria*, in Philipp Melanchthon, *Selectissimae orationes*, fol. 187v. Cf. Giovanni Pico della Mirandola, *Opera Omnia*, pp. 314–15, vol. I (italics mine): 'Igitur [Deus] hominem accepit indiscretae opus imaginis, atque in mundi positum meditullio, sic est alloquutus: "Nec certam sedem, nec propriam faciem, nec munus ullum peculiare tibi dedimus o Adam, ut quem sedem, quam faciem, quae munera tute optaveris, ea pro voto, pro tua sententia, habeas et possideas. Definita caeteris [animalibus] natura intra praescriptas a nobis leges coercetur: Tu nullis angustiis coercitus, pro tuo arbitrio, in cuius manu te posui, tibi illam praefinies. *Medium te mundi posui, ut circumspiceres inde commodius quicquid est in mundo*. Nec te coelestem, neque terrenum, neque mortalem, neque immortalem fecimus, ut turpi ipsius quasi arbitrarius honorariusque plastes et fictor, in quam malueris tute formam effingas. Poteris in inferiora quae sunt bruta degenerare. *Poteris in superiora quae sunt divina ex tui animi sententia regenerari*." O summam Dei patris liberalitetem, summam et admirandam hominis felicitatem. Cui datum id habere quod optat, id esse quod velit'.
I discuss the endurance of the classical origins and the endurance of an ethical conception of astronomy in Omodeo, *Copernicus in the Cultural Debates of the Renaissance*, pp. 323–31.

According to Regiomontanus, the difficulty of astronomy was increased by a deplorable contingency: many of the scientific works available in Latin were not translated in a way that could maintain the decorum and the beauty of the originals. Regiomontanus's adherence to typical Renaissance values is visible through his terminological choices, specifically by his treatment of qualities such as *decorum*, *facilitas*, and *facundias* of Ptolemy's Greek as opposed to the *oratio scabra et turbulenta* of medieval Latin. This demonstrates a commitment to stylistic harmony and formal perfection, which cannot but imply an ethical and cultural renaissance. Hence, his effort to foster the acquisition of the *Almagest* is part of a wider project to restore the knowledge and virtue of the ancestors:

> Hinc maior huius disciplinae [astronomicae] pendet difficultas: quamquam nec illa quidem parva est quae nobis peregrinorum codicum interpretes peperunt. Difficile est sane atque arduum, ut quae in aliena lingua bene dicta sunt: eundem decorem eandemque facilitatem in translatione conservent, etiam si a disertis viris summo studio ac diligentia translata fuerint. Ut si parum vel disertus vel curiosus fuerit interpres, scabra prorsus et turbulenta redundat oratio. Quod mihi plane evenisse videtur in praeclarissimo illo Ptolemei libro, quem *Magnam Compositionem* vocant, quia apud graecos mira facilitate facundiaque resplendeat, ita apud latinos durum ineptumque habetur, ut ne Ptolemeus quidem ipse si reviviscat ipsum sit pro suo recepturus.
>
> > (There is one major difficulty in this discipline [astronomy]. Although the translators of foreign codices procured us many works, nonetheless it is difficult and hard to preserve in a translation the same beauty and clarity that marks the sentences so well expressed in a foreign language, even though eloquent scholars accomplished their translations with great efforts and diligence. Their style looks very rough and turbulent as if the translators were either inarticulate or careless. It seems to me that this was the case with Ptolemy's most excellent work known as *Great Composition*. Whereas it shines in Greek for its clarity and fluency, it sounds obscure and inadequate in Latin so that, if Ptolemy came back to life, he would not recognize it as his own work.)[29]

This rejection of obscure medieval prose entails a moral judgement, since linguistic barbarism connotes those who speak inadequately. In this respect, it would be appropriate to mention Eugenio Garin's remark concerning the humanistic polemics against the obscurity of Scholastic philosophy: 'The condemnation of a style becomes the condemnation of a speculative approach'.[30] Additionally, in the Padua oration, Regiomontanus pitted the uncertainty of philosophical truths against the certainty of mathematical demonstrations, in

---

29 Johannes Regiomontanus, *Epytoma in Almagestum Ptolomei*; my translation.

30 Garin, 'Alle origini rinascimentali del concetto di filosofia scolastica', p. 472: 'La condanna di uno stile si fa condanna di una impostazione speculativa'.

particular the durability and never-ending validity of Euclid's theorems and Archimedes' inventions, which would last 'for the next thousand centuries' (*post mille secula venturis*). In contrast, the diverging Aristotelian schools of his time indicated a profound crisis in philosophical thought: '[...] neque Aristoteles ipse si reviviscеret, discipulos suos atque sequaces satis intelligere crederetur' ([...] I do not believe that even Aristotle would clearly understand his disciples and followers, if he came back to life).[31]

The Editorial Programme

According to Regiomontanus, the advance of mathematical studies should be achieved through the appropriation of the main classical sources (as mentioned in the dedication of the *Epytoma in Almagestum Ptolemaei*), as well as through a curricular university reform (as outlined in the *Oratio introductoria in omnes scientias mathematicas*). In the 1470s he gave a 'modern' dimension to his plan by inaugurating a printing house in Nuremberg. Reinhold later explained Regiomontanus's decision to move there as follows:

> Regiomontanus tranquillam sedem studiis suis quaerens anno Millesimo quadringesimo septuagesimo primo Noribergam se confert. In hac urbe ut mallet reliquum suae aetatis consumere, ipse has causas recitabat, quod artifices ibi essent idonei ad fabricandas machinas huic arti necessarias, deinde quod per mercatores facilius in alias regiones inde literas transmittere posset. Cum enim iam praeclare instructus esset doctrina, et caeterorum iudicia explorasset, et se quantum potuit, ad mathemata illustranda et propaganda praeparasset, constituerat et veteres Scriptores edere, et suas enarrationes publicare. Ad hanc rem instruxit officinam typographicam et nomenclaturam proposuit librorum, quos editurus erat, quae edhuc extat.
>
> (In 1471 Regiomontanus moved to Nuremberg in the search of a quiet place for his studies. He gave the following reasons for his decision to spend the rest of his life in this town: first, that the local craftsmen were capable of constructing the instruments [*machinas*] that are necessary to this art; and second, that it was easy to transmit letters through the traders. He had already been imparted an excellent education and had sufficiently explored the opinions of the others. Now he prepared himself to concentrate his energies in explaining and disseminating mathematics and planned to prepare editions of ancient authors and to publish his own treatises. For this purpose, he established a printing house and gave out a list of the books he was going to edit. The latter still exists.)[32]

31 Johannes Regiomontanus, *Oratio introductoria*, in Philipp Melanchthon, *Selectissimae orationes*, fol. 185^{v}; my translation.

32 Erasmus Reinhold, 'Oratio de Iohanne Regiomontano Mathematico', fol. 165; my translation.

The list Reinhold mentions still exists today. In 1474 Regiomontanus issued this famous broadside, *Haec opera fient in oppido Nuremberga Germania ductu Ioannis de Monteregio* (These Works Will Be Printed by Johannes Regiomontanus in the Town of Nuremberg, Germany), presenting his trade list. The titles he intended to print comprised many works from classical antiquity (among them works by Euclid, Archimedes, Theodosius and Ptolemy), his own manuscripts and less conspicuously, medieval and recent works such as Witelo's optics, Jordanus Nemorarius' *Arithmetica* and Peuerbach's *Theoricae novae planetarum* (in fact, printed in 1474). In Rose's eyes, 'in its fusion of mathematics, Greek and printing Regiomontanus's publishing Programme of 1474 marks the formal beginning of the renaissance of mathematics'.[33]

Although Regiomontanus was hindered from finishing his plans by an untimely death, his project inspired many new publications. His pupils, especially Johannes Schöner (1477–1547), continued the project of publishing ancient and new works on mathematical and natural sciences, and the publisher Johannes Petreius (*c.* 1497–1550), initiated a series comprised of titles from Regiomontanus's trade list (such as Regiomontanus's *De triangulis* in 1533 and Witelo's *Optics* in 1535). Among many noteworthy publications, Petreius also published Copernicus's groundbreaking *De revolutionibus orbium coelestium* (1543) and Girolamo Cardano's works on astrology (*Libelli duo, unus de supplemento Almanach, alter de Restitutione temporum et motuum cœlestium, item geniturae LXVII*, 1543), algebra (*Ars magna*, 1545), and his highly controversial *De subtilitate* (On Subtlety) (1550).[34]

Nuremberg: A Crossroads

In 1537 Petreius printed Regiomontanus's oration as part of an astronomical miscellanea, of which the most extensive works included two Arabic astronomical treatises discussing spherical astronomy and planetary theory. The first astronomical source was Alfarghani's *Rudimenta astronomica* (Elements of Astronomy), in John of Seville's (died 1166) translation, and the second source was Albategnius's *De motu stellarum*, in Plato of Tivoli's translation. Johannes Schöner took upon himself the responsibility of revising and editing the texts. The original source of *De motu stellarum* came from Regiomontanus's library, whereas the edition of Alfarghani's 'sphere' was probably derived from a manuscript different from the one owned by Regiomontanus.[35] In addition to Regiomontanus's *Oratio introductoria in omnes scientias mathematicas*, the

33 Rose, *The Italian Renaissance of Mathematics*, p. 110.

34 For an overview of the publication of the books listed in Regiomontanus's Program, see Malpangotto, *Regiomontano e il rinnovamento del sapere matematico e astronomico nel Quattrocento*, pp. 211–17.

35 Cf. Gingerich, 'Reinhold, Erasmus'.

miscellanea also included his introduction to Euclid's *Elements*. Moreover, it began with a dedication written by Melanchthon to the Nuremberg Senate, which made clear that this editorial endeavour was part of a Lutheran-humanistic programme of religious and scholarly reformation.[36]

Melanchthon's letter is explicitly political. The educational reformer reminds the Nuremberg patricians of the importance of cultivating the arts alongside military discipline in a well-ruled republic. In this light, Melanchthon praises the allocation of a stipend to Schöner as town mathematician and mentions the fact that Regiomontanus made Nuremberg his hometown. He then begins a discussion on the different forms of government, expressing uncertainty as to whether a monarchy or an aristocratic regime would be preferable (*vetus disputatio est de Monarchia et Aristocratia*). In any case, he argues that aristocracy serves to temper monarchy in order to avoid tyranny. As for democracy, he contends that it is a dangerous form of anarchy, in which all harmony is destroyed and the plebs take over power. In summary, democracy is a political model that counters justice, as is allegedly revealed by mathematics:

> Itaque et Aristoteles eam iusticiam, quae imperia constituit, imitari proportionem Geometricam dixit, quae dissimilis est regno et Democratiae. Nam proportio multos inter se consociat, et grauds constituit, ne promiscue rapiant imperia infimi.
>
> > (According to Aristotle, the justice that holds together a state [*imperium*] imitates geometrical proportion, unlike the case of democracy. In fact, proportion permits the association of many people while establishing hierarchies [*gradus*] so that the plebs do not bring disorder by vulgarly thieving power [*imperium*].)[37]

This publication of 1537 marks the insertion of Regiomontanus's humanist programme in the Lutheran-Philippist Reformation agenda. Later, in his public celebration of Regiomontanus of 1549, Reinhold reminded Wittenberg scholars that the editorial legacy of Regiomontanus was still alive in Nuremberg:

> Bibliothecam vero Regiomontani Walterus ille civis Norimbergensis [...] redemit, quo mortuo senatus adservari reliquias curavit, ex quibus vir

36 Melanchthon's ties to the mathematical culture in Nuremberg, in particular with his friend and protégé Johannes Schöner has been stressed by Reich and Knobloch, 'Eine besondere Freundschaft'; see also Reich, 'Philipp Melanchthon im Dialog mit Astronomen und Mathematikern'.

37 Melanchthon, *Epistola nuncupatoria ad Senatum Norimbergensem*, in Alfraganus, *Rudimenta astronomica Alfragani item Albategnius*, fol. a2^{v}; my translation. The passage continues as follows: 'Tantamque vim huius proportionis esse Plato scripsit, ut quidquid boni contingit civitatibus, hanc proportionem esse causam affirmet, quae cum ad Aristocratiam proprie quadret, philosophi hunc statum anteferunt. Porro non facile usque meliorem huius status effigiem videre liceat, quam in vestra Repub[lica].'

doctissimus Schonerus quaedam utilia scripta edidit. Sed plura alia et longe utiliora haberemus, nisi tristo fato author ante tempus extinctus esset.

> (The famous Nuremberg citizen, Walther bought Regiomontanus's library [...]. On his death, the Senate looked after the relics, out of which the very learned Schöner edited some useful writings. Yet, we would have many more and by far more useful ones, had evil fate not taken away that author too early.)[38]

As far as Regiomontanus's editorial legacy was concerned, Nuremberg was the place designated to continue it. As for the academic legacy, Reinhold suggested that Wittenberg was the predestined centre where mathematics could advance under Melanchthon's auspices.

Wittenberg 1549: Accomplished Reformation

The *Oratio de Iohanne Regiomontano*, which Reinhold delivered in 1549 to mark the occasion of the awarding of eight Master's degrees, was integrated into a ritualized institutional setting. As he declared to his listeners, the occurrence of such official academic gatherings (*in his scholasticis congressibus*) required the participants to bring together prayers (*vota*) and learned discourses (*commonefactiones de doctrina*). This established tradition (*cum usitatum sit...*) obliged him as an orator to celebrate a well-known figure of the past, an erudite scholar in particular, so as to inspire the audience to follow such an example. As a reformed professor of mathematics who was trained in humanism, Reinhold regarded Regiomontanus as endowed with all the relevant academic values, among them, the force and sharpness of ingenuity (*ingenii vis et acies*), erudition, and correct judgement (*iudicii rectitudo*). Reinhold presented Regiomontanus as a *clericus vagans* who lived in Germany, Italy, and Hungary; who mastered two 'foreign languages', Greek and mathematics; and who also proved to be a skillful craftsman in constructing instruments and launching his own typography. Reinhold repeats several times in this speech that this model ought to be followed: 'Regiomontanus imitemur'.[39]

Reinhold made astrology, seen as a pious contemplation of God's Providence, the main focus of his talk. This can be regarded as typical of the so-called 'Wittenberg School'. In the dedicatory epistle of the *Rudimenta astronomica Alfragani item Albategnius* (1537), Melanchthon stressed that astronomical studies can dispel atheist views by revealing God's hidden

38 Erasmus Reinhold, 'Oratio de Iohanne Regiomontano Mathematico', fol. 169^r.

39 Erasmus Reinhold, 'Oratio de Iohanne Regiomontano Mathematico', fol. 169^r; see also fol. 160^r, and fol. 169^v: 'Imitemur [...], quantum possumus, eius studia et sedulitatem. Denique has ipsas artes, quas illustravit, amemus et magnificiamus'.

design and refuting the Epicurean notion that nature is ruled by chance.[40] Reinhold also displayed his bias toward astrology by discussing in the *Oratio de Regiomontano*, Regiomontanus's horoscope (printed for the first time by Petreius in 1547 in Cardano's collection of one hundred genitures, *Liber de exemplis centum geniturarum*).[41] Moreover, he wrote that Regiomontanus used astrology to cure King Mathias Corvinus of Hungary, proving a better healer than the court physicians.[42] Reinhold also claimed that by means of astrology, Regiomontanus foresaw that travelling to Rome to serve the pope and implement the calendar reform would be fatal; nevertheless he embarked on this trip in 1475 on behalf of the common good.

In Reinhold's eyes, Regiomontanus represented a learned scholar who tirelessly worked toward the restoration of mathematical and classical studies, which were seen as two faces of the same coin. However, his historical perspective was quite different than Regiomontanus's. For Reinhold, only classical and biblical figures possessed a paradigmatic status. Latin and Arabic medieval authors were downgraded to a lesser status. Reinhold illustrated the master-pupil relation between Peuerbach and Regiomontanus by comparing cases of ancient thinkers such as Thales, Meton, and Eudoxus, all educated by Egyptian priests, with emphasis on Eudoxus, who was a pupil of Plato. In order to illustrate the importance of teaching, he resorted to mythical figures such as Atlas, who taught Orion astronomy and to Biblical traditions such as the transmission of astronomical knowledge along the path connecting Adam to Noah and their descendants. The Hebrews were considered to have brought astronomy to Egypt, where it was cultivated and flourished up to the time of Ptolemy.[43] As for the Arabs, Reinhold did not praise them as a source of medieval mathematical knowledge in the Latin West but rather as the cause of its corruption. Peuerbach and Regiomontanus's *Epytoma* appears to him as an almost miraculous *generatio spontanea*:

> Bessarion vir Graecus doctrina excellens, cum esset Cardinalis, tunc ad Fridericum imperatorem legatus erat. Is cum de Ptolemaei doctrina Purbacchium multa sciscitatus esset, et dexteritatem eius in explicando admiraretur, petivit ab eo, ut Epitomen componeret eius libri, cui titulus est μεγάλη σύνταξις. Id opus interruptu morte Purbacchii Regiomontanus

40 Philipp Melanchthon, *Epistola nuncupatoria ad Senatum Norimbergensem*, in Alfraganus, *Rudimenta astronomica Alfragani item Albategnius*, fol. a3^{v}: 'Nam si Epicuri deliramenta defendere conantur, qui furenter contendebat coelestia corpora nulla certa lege, sed casu ferri, digni odio sunt, vel stigmatis potius, quod adolescentes a recte et verae philosophiae studiis abducunt'. On Melanchthon's anti-Epicureanism and his legacy in the astronomical culture of the Renaissance, cf. Boner, 'Kepler's Copernican Campaign and the New Star of 1604'.

41 Girolamo Cardano, *De Exemplis Centum Geniturarum*, p. 498; cf. Malpangotto, *Regiomontano e il rinnovamento del sapere matematico e astronomico nel Quattrocento*, p. 25 n. 12.

42 Erasmus Reinhold, 'Oratio de Iohanne Regiomontano Mathematico', fol. 164^{r-v}.

43 Erasmus Reinhold, 'Oratio de Iohanne Regiomontano Mathematico', fol. 161^{r}.

> absolvit […]. Qua in re valde mirandum est, tunc sine Graecis fontibus sententiam eos *ex illis Arabicae versionis tenebris* eruere potuisse.
>
> (When he was a Cardinal, the very learned Greek Bessarion was sent as a papal legate to the Emperor Frederick. He asked many questions to Peuerbach on Ptolemy's doctrine and was very admired of the latter's capacity of explaining it and therefore required him to compose an *Epitome* of [Ptolemy's] book entitled μεγάλη σύνταξις. After this endeavour was interrupted owing to Peuerbach's death, Regiomontanus completed it […]. It is indeed a wonder that they could explain the meaning [of the *Almagest*] without Greek sources and only relying on *the obscurity of Arabic versions*.)[44]

For Reinhold, the mastery of Greek counts as a valuable academic skill that must be flaunted. Greek words are used throughout his Latin oration in order to elevate its style. Ptolemy's major astronomical work is mentioned in Greek as μεγάλη σύνταξις and a long passage of Plato's *Laws* is quoted from the original.[45] Knowledge of Greek appears to be indispensable as the key that opens the philosophical treasures of classical science. Regiomontanus, for one, is acknowledged as somebody who not only transcribed Greek and Latin books by hand but also decorated them.[46] Actually, he did not limit himself to mathematics and philosophy. As Reinhold reports,

> Novum Testamentum integrum graece sua manu elegantissime descripsit, quem Codicem mihi Schonerus inter alia eius monumenta monstravit.
>
> (He transcribed the entire New Testament in Greek in his own hand in a very elegant manner. Schöner showed this codex to me alongside other of his works.)[47]

Piety was a virtue connected with erudition. Regiomontanus's achievements were deemed impossible without God's grace (*sine adflatu divino*).[48] The last part of Reinhold's speech is centred on the theological meaning of scholarship inspired by Regiomontanus:

> Et iam lectio Epistolae Paulinae, quae hodie publice recitatur, ut mos est Ecclesiae, illustre testimonium est, studia linguarum et verae Philosophiae necessaria esse ad explicationem monumentorum, in quibus immensa bonitate Deus extare voluit doctrinam de suo filio, et de reconciliatione.

---

44 Erasmus Reinhold, 'Oratio de Iohanne Regiomontano Mathematico', fols 161v–162r (italics mine); my translation.

45 Erasmus Reinhold, 'Oratio de Iohanne Regiomontano Mathematico', fol. 162r and fol. 160r.

46 Erasmus Reinhold, 'Oratio de Iohanne Regiomontano Mathematico', fols 163v–164r.

47 Erasmus Reinhold, 'Oratio de Iohanne Regiomontano Mathematico', fol. 168r; my translation.

48 Erasmus Reinhold, 'Oratio de Iohanne Regiomontano Mathematico', fol. 159v.

> (Today's lecture of Paul's epistle, publicly held according to the custom of our Church, clearly bears witness that the study of languages and of true philosophy is necessary for the explanation of the monuments in which God, in his immense goodness, wanted to reveal the doctrine concerning his son and salvation.)[49]

Study must be directed *ad maiorem gloriam Dei*.[50] In this life, as Reinhold concludes, we should cultivate the sciences in order to be prepared for the eternal university (*aeterna schola*) of prophets and apostles. In this manner, Regiomontanus's renaissance of mathematics is transcended on the level of a theological and ecclesiastical reformation. In 1551 the *Oratio de Iohanne Regiomontano* and Regiomontanus's *Oratio… quam habuit ipse Patavii in praelectione Alfragani* in praise of mathematics were both reprinted and included in the third tome of Melanchthon's orations. The first of them is an *Oratio in funere Reverendi viri D. Martini Lutheri* (Funerary Oration for Reverend Martin Luther). This Lutheran re-contextualization of the work of the fifteenth-century mathematician could not be more clearly stressed.

Summary

In 1549, when Reinhold publicly presented Regiomontanus as a model of scholarly virtues and pious erudition, the Latin mathematical culture had greatly changed relative to the humanistic culture of the earlier century. The publication of Copernicus's major work in 1543 made German scholars of the Melanchthonian milieu believe that Latin scholarship, especially in Germany, had reached and perhaps even surpassed the level of its predecessors. In a certain sense, mathematical astronomers such as Reinhold could be persuaded that the renaissance of mathematics envisioned by Regiomontanus had been accomplished. As we have seen, Regiomontanus's programme had various facets besides the acquisition of certain standards of mathematical preparation. First, it aimed to recover and translate classical sources; second, it aimed to introduce the teaching of these sources at universities, and third, it envisaged their publication alongside new works in mathematics. As far as the original programme by Regiomontanus is concerned, one could say that in the Wittenberg reception the first point, the rediscovery of classical sources, was perceived either as completed or coming to an end. As for the second point, one could say that it had been realized by the Melanchthonian university reform, in which mathematical

49 Erasmus Reinhold, 'Oratio de Iohanne Regiomontano Mathematico', fol. 170r; my translation.

50 Erasmus Reinhold, 'Oratio de Iohanne Regiomontano Mathematico', fol. 175r: 'Ideo vos propter Deum et propter Ecclesiam obtestamur, ut studia Philosophica ametis, ad quorum conservationem variae operae scholasticae conducunt'.

disciplines received an unprecedented relevance. Third, the publication programme had marched on, especially in Nuremberg, where scholars such as Schöner and publishers such as Petreius ideally continued his activity. A fourth and a fifth element were added in the wake of the Wittenberg reformation. One encounters an explicit emphasis on the political meaning of mathematical disciplines in connection with moral and pious behaviour. Finally, the Renaissance idea of a scholarly renewal was absorbed by the Reformation project of a pious form of knowledge directly linked to the sources, bypassing tradition. Thus, the shift from Regiomontanus to Reinhold could perhaps be described as a shift from Renaissance to Reformation. It implied a change in historical perspectives. The most evident is a historical reassessment according to which classical and biblical sources became the privileged point of reference for science and theology at the expense of the acknowledgement of the transcultural exchanges that had taken place since the Middle Ages.[51]

Bibliography

Primary Sources

Alfraganus [al-Farghānī], *Rudimenta astronomica Alfragani item Albategnius* (Nuremberg: Petreius, 1537)

Erasmus Reinhold, 'Oratio de Iohanne Regiomontano Mathematico', in *Selectissimarum orationum clarissimi viri domini Philippi Melanchthonis*, III (Erfurt: Excussit Gervasius Sturmer, 1551), fols 157r–172v

Giovanni Pico della Mirandola, *Opera Omnia* (Basel: per Heinricum Petri, 1557)

Girolamo Cardano, *De Exemplis Centum Geniturarum* (Nuremberg: Petreius, 1547)

Johannes Regiomontanus, *Epytoma in Almagestum Ptolomei* (Venezia: Iohannes Ha[m]man de Landoia, dictus Hertzog, 1496)

———, *In Elementa Euclidis Praefatio*, in Alfraganus, *Rudimenta astronomica Alfragani item Albategnius* (Nuremberg: Petreius, 1537)

Nicolaus Copernicus, *Gesamtausgabe (Receptio Copernicana)*, vol. VIII, 1 (Hildesheim: Gerstenberg, 2002)

Philipp Melanchthon, *Selectissimae orationes*, III (Erfurt: Excussit Gervasius Sturmer, 1551)

51 Cf. Brentjes, Fidora, and Tischler, 'Towards a New Approach to Medieval Cross-Cultural Exchanges'.

Secondary Studies

Boner, Patrick, 'Kepler's Copernican Campaign and the New Star of 1604', in *Change and Continuity in Early Modern Cosmology*, ed. by Patrick Boner (Dordrecht: Springer, 2011), pp. 93–114

Brentjes, Sonja, Alexander Fidora, and Matthias M. Tischler, 'Towards a New Approach to Medieval Cross-Cultural Exchanges', *Journal of Transcultural Medieval Studies*, 1.1 (2014), 9–50

Brosseder, Claudia, *Im Bann der Sterne: Caspar Peucer, Philipp Melanchthon und andere Wittenberger Astrologen* (Berlin: de Gruyter, 2004)

Byrne, James Steven, 'A Humanist History of Mathematics? Regiomontanus's Padua Oration in Context', *Journal of The History of Ideas*, 67.1 (2006), 41–61

Cantor, Moritz, *Vorlesungen über Geschichte der Mathematik (1200–1668)* (Leipzig: Teubner, 1892)

Garin, Eugenio, 'Alle origini rinascimentali del concetto di filosofia scolastica', in *La cultura filosofica del rinascimento italiano* (Milano: Bompiani, 1994)

Gingerich, Owen, 'Reinhold, Erasmus', *Dictionary of Scientific Biography*, 11 (1975), 365–67

Goulding, Robert, *Defending Hypatia: Ramus, Saville, and the Renaissance Rediscovery of Mathematical History* (Dordrecht: Springer, 2010)

Hotson, Howard, *Commonplace Learning: Ramism and Its German Ramifications 1543–1630* (Oxford: Oxford University Press, 2007)

Jardine, Nicholas, 'Rothmann's Dialexis on the Comet of 1585', in *Christoph Rothmann's Discourse on the Comet of 1585*, ed. by Miguel Ángel Granada, Adam Mosley, and Nicholas Jardine (Leiden: Brill, 2014), pp. 258–81

Kusukawa, Sachiko, *The Transformation of Natural Philosophy: The Case of Philip Melanchthon* (Cambridge: Cambridge University Press, 1995)

Malpangotto, Michela, *Regiomontano e il rinnovamento del sapere matematico e astronomico nel Quattrocento* (Bari: Caucci, 2008)

Omodeo, Pietro Daniel, *Copernicus in the Cultural Debates of the Renaissance: Reception, Legacy, Transformation* (Leiden: Brill, 2014)

———, 'Liddel's Oratio de Praestantia Mathematicarum', in *Duncan Liddel (1561–1613), Networks of Polymathy and the Northern European Renaissance*, ed. by Pietro Daniel Omodeo, with Karin Friedrich (Leiden: Brill, 2016), pp. 218–36

———, 'Utilitas Astronomiae in the Renaissance: Rhetoric and Epistemology of Astronomy', in *The Structures of Practical Knowledge*, ed. by Matteo Valleriani (Cham: Springer, 2017), pp. 307–32

Reich, Karin, 'Philipp Melanchthon im Dialog mit Astronomen und Mathematikern: Ausgewählte Beispiele', in *Mathematik und Naturwissenschaften in der Zeit von Philipp Melanchthon*, ed. by Franz Fuchs (Wiesbaden: Harrassowitz, 2012), pp. 27–58

Reich, Karin, and Eberhard Knobloch, 'Eine besondere Freundschaft: Johannes Schöner (1477–1547) und Philipp Melanchthon (1497–1560)', in *Lebensläufe und Himmelsbahnen*, ed. by Wolfgang Dick and Dietmar Fürst (Leipzig: Akademische Verlagsanstalt, 2014), pp. 15–52

Rose, Paul Lawrence, *The Italian Renaissance of Mathematics* (Geneva: Droz, 1975)
Scheible, Heinz, *Melanchthon in seinen Schülern* (Wiesbaden: Harrassowitz, 1997)
Shank, Michael H., 'Regiomontanus on Ptolemy, Physical Orbs, and Astronomical Fictionalism: Goldsteinian Themes in the "Defence of Theon against George of Trebizond"', *Perspectives on Science*, 10.2 (2002), 179–207
Thüringer, Walter, 'Paul Eber (1511–1569): Melanchthons Physik und seine Stellung zu Copernicus', in *Melanchthon in seinen Schülern*, ed. by Heinz Scheible (Wiesbaden: Harrassowitz, 1997), pp. 285–321
Zinner, Ernst, *Regiomontanus: His Life and Work* (Amsterdam: North-Holland, 1990)

Index of Premodern Actors

Contact and Transmission

Intercultural Encounters from Late Antiquity to the Early Modern Period

All volumes in this series are evaluated by an Editorial Board, strictly on academic grounds, based on reports prepared by referees who have been commissioned by virtue of their specialism in the appropriate field. The Board ensures that the screening is done independently and without conflicts of interest. The definitive texts supplied by authors are also subject to review by the Board before being approved for publication. Further, the volumes are copyedited to conform to the publisher's stylebook and to the best international academic standards in the field.

Titles in Series

Isaac Lampurlanés Farré, *Excerptum de Talmud: Study and Edition of a Thirteenth-Century Latin Translation* (2020)

In Preparation

Narratives on Translations across Eurasia and Africa: From Antiquity to Modern Times, ed. by Sonja Brentjes, Jens Høyrup, and Bruce R. O'Brien